The Optics of Giambattista Della Porta (ca. 1535–1615): A Reassessment

Archimedes

NEW STUDIES IN THE HISTORY AND PHILOSOPHY OF SCIENCE AND TECHNOLOGY

VOLUME 44

Archimedes has three fundamental goals; to further the integration of the histories of science and technology with one another: to investigate the technical, social and practical histories of specific developments in science and technology; and finally, where possible and desirable, to bring the histories of science and technology into closer contact with the philosophy of science. To these ends, each volume will have its own theme and title and will be planned by one or more members of the Advisory Board in consultation with the editor. Although the volumes have specific themes, the series itself will not be limited to one or even to a few particular areas. Its subjects include any of the sciences, ranging from biology through physics, all aspects of technology, broadly construed, as well as historically-engaged philosophy of science or technology. Taken as a whole, *Archimedes* will be of interest to historians, philosophers, and scientists, as well as to those in business and industry who seek to understand how science and industry have come to be so strongly linked.

More information about this series at http://www.springer.com/series/5644

Arianna Borrelli • Giora Hon • Yaakov Zik
Editors

The Optics of Giambattista Della Porta (ca. 1535–1615): A Reassessment

Springer

Editors
Arianna Borrelli
History of Science
Technische Universität Berlin
Berlin, Germany

Institute for Advanced Study on "Media Cultures of Computer Simulation (MECS)"
Leuphana University
Lüneburg, Germany

Giora Hon
Department of Philosophy
University of Haifa
Haifa, Israel

Yaakov Zik
Department of Philosophy
University of Haifa
Haifa, Israel

ISSN 1385-0180 ISSN 2215-0064 (electronic)
Archimedes
ISBN 978-3-319-50214-4 ISBN 978-3-319-50215-1 (eBook)
DOI 10.1007/978-3-319-50215-1

Library of Congress Control Number: 2017932422

Printed on acid-free paper

This Springer imprint is published by Springer Nature
The registered company is Springer International Publishing AG
The registered company address is: Gewerbestrasse 11, 6330 Cham, Switzerland

Contents

1 **Introduction** 1
Arianna Borrelli

2 **A Theater of Experiments: Giambattista Della Porta and the Scientific Culture of Renaissance Naples** 11
William Eamon

3 **Giambattista Della Porta: A Magician or an Optician?** 39
Yaakov Zik and Giora Hon

4 **Optical Diagrams as "Paper Tools": Della Porta's Analysis of Biconvex Lenses from *De refractione* to *De telescopio*** 57
Arianna Borrelli

5 **Giambattista Della Porta's Theory of Vision in the *De refractione* of 1593: Sources, Problems, Implications** 97
A. Mark Smith

6 **Magi from the North: Instruments of Fire and Light in the Early Seventeenth Century** 125
Fokko Jan Dijksterhuis

7 **Using Invariances in Geometrical Diagrams: Della Porta, Kepler and Descartes on Refraction** 145
Albrecht Heeffer

8 **Francesco Maurolico, Giambattista Della Porta and Their Theories on Refraction** 169
Riccardo Bellé

9 **Conclusion: A Reassessment** 201
Yaakov Zik and Giora Hon

Chapter 1
Introduction

Arianna Borrelli

Giambattista Della Porta (ca. 1535–1615) was an outstanding representative of the vitality and variety of Southern Italian culture in the late Renaissance, but at the same time his work displayed features resonating with the many approaches to the experimental philosophy of nature being developed all over Europe.[1] Already in the early years of his long life Della Porta achieved fame at European level with his "Four books on natural magic" ("Magiae naturalis libri quatuor"), published in 1558 and expanded into a twenty-book edition in 1589.[2] The "Natural magic" was a collection of experiments on a very broad range of subjects, which were presented in a form appealing to a general audience. Later on, Della Porta took up and expanded upon many of these areas of knowledge in thematic monographs directed at a more learned public. This was the case also with optics, a discipline that in the early modern period underwent a fundamental transformation during which three general areas of natural inquiry which had traditionally been closely linked eventually came to be separated: visual perception and the physiology of the eye, the nature and properties of light, and the geometrical rules of perspective, reflection and refraction.[3]

In Della Porta's optical writings the distinctions among these three areas did not yet fully obtain, but were already underway. His works are of particular interest for

[1] On Della Porta's life and cultural context see William Eamon's contribution to the present volume, which also contains relevant references to sources and secondary literature.

[2] Della Porta 1558, 1589.

[3] For an overview on the development of pre-modern and early modern optics see: Lindberg 1976, Smith 2015.

A. Borrelli (✉)
History of Science, Technische Universität Berlin, Berlin, Germany

Institute for Advanced Study on "Media Cultures of Computer Simulation (MECS)", Leuphana University, Lüneburg, Germany
e-mail: aborrelli@weatherglass.de

A. Borrelli et al. (eds.), *The Optics of Giambattista Della Porta (ca. 1535–1615): A Reassessment*, Archimedes 44, DOI 10.1007/978-3-319-50215-1_1

the history of early modern optics since they represent the first extended and systematic textual sources on the optical artifacts which became available in increasing numbers during the sixteenth-century thanks to the progress in glass-making and glass-working. These developments, and in particular the rise of the epistemic significance of the telescope, played a key role in leading early modern authors to deal separately with the physiology of vision, the theory of light and geometrical optics.

Many references to Della Porta's optical work were made in context of the dispute on the invention of the telescope, a discussion in which Della Porta himself was engaged in the last years of his life. In light of the huge amount of literature on this subject, little if anything remains to be said, and the priority discussion plays no role in the present volume, although the telescope will feature in Fokko Jan Dijksterhuis' contribution on the reception of Della Porta's work in the Netherlands. The studies collected here focus on Della Porta's published and unpublished attempts at making sense of optical phenomena made possible by the newly developed glass artifacts, and aim at reassessing their original content and their role in the development of early modern optics. Della Porta discussed optical themes in four works:

- chapter 1 to 19 of book IV of the first edition of "Natural magic" (1558)[4]
- the whole book XVII of the second edition of "Natural magic" (1589)[5]
- the treatise "On refraction" ("De refractione", 1593)[6]
- the manuscript draft of the unfinished treatise "On the telescope" ("De telescopio", ca. 1610–1615)[7]

Della Porta's writings in general and his optical works in particular have not been treated kindly by modern historiography, and he has often been portrayed either as a chaotic, enthusiastic experimenter hopelessly stuck in a medieval worldview, or as a talented showman only interested in amazing and entertaining naive audiences rather than in furthering the understanding of nature.[8] To understand the motivations for a reassessment of Della Porta's optics it is important to look back at how historians have dealt with it, and this is what I will do in the following pages. The overview does not aim at exhaustiveness, but only at sketching how the present, mainly negative image of the Neapolitan scholar emerged.

Della Porta's contemporaries had a rather high estimate of his optical research: Johannes Kepler is known to have read carefully the optical experiments in the "Natural magic" and sought for a copy of the treatise "On refraction", which Thomas Harriot used as a basis for his own optical work.[9] A relatively long and positive discussion of Della Porta's optical writings is found in Joseph Priestley's "History

[4] Della Porta 1558.

[5] Della Porta 1589.

[6] Della Porta 1593.

[7] Della Porta 1962.

[8] As examples of such views, see the rather old entry on Della Porta in the Dictionary of Scientific Biography (Rienstra 1975) and the more recent one in the Stanford Encyclopaedia of Philosophy (Kodera 2015).

[9] Duprè 2008, Goulding 2013.

and present state of discoveries relating to vision, light and colour" (1772), and shorter mentions appeared later in Emil Wilde's "History of optics" ("Geschichte der Optik" 1838), in Raffaello Caverni's "History of the experimental method in Italy" ("Storia del metodo sperimentale in Italia" 1891), and in Edmund Hoppe's "History of optics" ("Geschichte der Optik" 1926).[10] However, the person who most extensively studied Della Porta's optical work was also one of his harshest critics: Vasco Ronchi, eminent scientist and expert of optics, as well as historian of this discipline. In various publications Ronchi discussed in some detail Della Porta's treatise "On refraction"; he also took care that the manuscript draft of the treatise "On the telescope", discovered in 1940 in the archive of the Accademia dei Lincei, should be edited with great competence by Maria Amalia Naldoni.[11] In a paper published in 1947 Ronchi gave factually accurate discussions of Della Porta's treatment of lenses and stated clearly that all phenomena presented by Della Porta were empirically correct. However, although he fully acknowledged Della Porta's experimental capabilities, Ronchi missed in Della Porta's writing a theory of lenses starting from first principles. According to Ronchi, Della Porta's ray-tracing constructions led to correct empirical conclusions, but they did not prove those conclusions on the basis of some fundamental principles:

> [Della Porta] belongs to the number of those, and even today there are many, who do not give any importance to the definition of fundamental concepts, a task that they are happy to leave to the philosophers, and instead start from any point of departure, as long as it fit for their operations.[12]

After describing one of Della Porta's constructions, Ronchi concluded:

> As we see, here there is anything but the proof. And the same goes for all other propositions, about a dozen, following the first one. But apart from the logic, there is a content, both an experimental and a theoretical one. The experimental one is more definite, because numerous and accurate observations must have been made. [...] The most interesting theoretical content consists in the attempt to define the position of the image given by the lens using the intersection of a ray with an orthogonal line, as in the case of plane refracting surfaces.[13]

Ronchi gave a positive assessment of the experimental content of Della Porta's work, but harshly criticized his theoretical methodology. The problem, for Ronchi, was not just that Della Porta did not arrive at the law of refraction as stated by later authors, but also that he was not even aiming at anything of the sort: Della Porta was

---

[10] Priestley 1772, 34–43; Wilde 1838 vol. 1, 112–12; Caverni 1891, 536–566; Hoppe 1926, 24–25.

[11] Della Porta 1962, Naldoni 1946, 1962, Ronchi 1947, 1962, 1970.

[12] "[Della Porta] si mette nel novero di quei non pochi, anche oggi, che non danno peso alla definizione dei concetti fondamentali, opera che essi lasciano ben volentieri ai filosofi, ma partono da una base qualunque, purchè vada bene per le loro operazioni." (Ronchi 1947, 11).

[13] "Come si vede, c'è tutto fuorchè la dimostrazione. E dello stesso tipo sono tutte le altre [proposizioni], circa una dozzina, che seguono questa prima. Ma a parte la logica, un contenuto vi è, sia sperimentale, sia teorico. Quello sperimentale è più definito, perchè le osservazioni fatte debbono essere state numerose e accurate.[...] Il contenuto teorico più interessante consiste nel tentativo di definire la posizione dell'immagine data dalla lente, ricorrendo all'intersezione di un raggio con una normale, come a proposito delle superfici rifrangenti piane" (Ronchi 1947, 13–14).

apparently formulating what today's scientists would call a "phenomenological theory" of lenses. All in all, Ronchi's discussion of Della Porta's work in his 1947 paper was less biased than in other works. For example, in his monograph on "The nature of light" (1970, original Italian edition 1939), a book which became well known at the international level, he skipped any praise for Della Porta's experimental results, and underscored in a more critical tone the (alleged) defects of his theoretical exposition, stating that Della Porta "was triumphantly writing what we can only call his absurdities on lenses and refraction".[14] Nonetheless, Ronchi in all his writings underscored how Della Porta, togehter with Francesco Maurolico, paved the way for Kepler's reflections, which for Ronchi represented the turning point from ancient to modern optics.[15] Ronchi remained for many decades the only historian who had directly engaged with Della Porta's optical writings, and his judgment led other scholars to formulate a negative opinion on the subject without delving into Della Porta's writings to the same extent as Ronchi had done. David Lindberg was the next author who looked more closely into Della Porta's optical work and he went even further than Ronchi with his critique. Lindberg discussed Della Porta in a paper aimed at disproving Ronchi's claim that Della Porta and Maurolico "introduced important innovations into optical theory, which prepared the way and laid the groundwork for Kepler's revolutionary activities in the seventeenth century".[16] Lindberg took up Ronchi's critique that Della Porta's geometrical constructions were no rigorous proofs, but did not even mention their empirical accuracy and systematic character; at times he even read incorrect interpretations into Della Porta's propositions.[17] Lindberg took issue with Ronchi's claim that Della Porta's failures were the failures of medieval optical theory, and noted that Kepler also followed those principles, but had still managed to develop "an adequate theory of thin lenses".[18] What is, or not, "an adequate theory" is however a question which depends not only on the theory, but also on the criteria of adequacy, and both Ronchi and Lindberg were using modern geometrical optics as a benchmark for their judgment. The contributions in this volume will instead offer a number of other perspectives from which the "adequacy" of Della Porta's optics can be assessed.

Resuming our brief historiographical overview, it is worth mentioning a paper by Johannes Lohne (1963) discussing the history of the law of refraction.[19] Lohne describes the rule for refraction in a glass sphere given in Della Porta's "On refraction" and characterizes it as "benchmark formula" ("Werkstattformel") which would have provided an estimate of the refraction angle very near to the modern one under the assumption that the glass employed had refracting index equal to 1,5.[20]

[14] Ronchi 1970, 86.

[15] Ronchi 1970, 77–109.

[16] Lindberg 1984, 131.

[17] Borrelli 2014, 50–51, 56.

[18] Lindberg 1984, 148.

[19] Lohne 1963.

[20] Lohne 1963, 157–158.

More recently, in a series of studies on Renaissance optics Sven Duprè has brought to light the contribution of artists and artisans to its development, and in this context he approached Della Porta's work with more appreciation of his practice-oriented methods than earlier historians. Still, his negative remarks on Della Porta's constructions are often based on Lindberg's incorrect interpretations.[21] A further historian who devoted much attention to Della Porta is Fumikazo Saito: In 2008 he discussed in detail Della Porta's theory of vision underscoring how, for him and his contemporaries, optics was the study of "how things are seen", and that their work has to be approached without comparing it to modern optics.[22] In his later monograph on the role of the telescope in Della Porta's natural magic, Saito discusses passages from all of Della Porta's published and unpublished optical writings, and expands on his previous reflections. He argues that, for Della Porta, geometry was a tool for formulating constructions fitting to experience, and not a means to deduce proofs from Euclidean principles.[23] According to Saito, Della Porta did not aim at theoretically understanding the functioning of lenses and telescopes, but at giving "the possibility of considering an instrument as mediator between experimenter and the object of experiment".[24] Saito's assessment of Della Porta is more positive than those of previous authors, but he shows little interest in analysing Della Porta's geometrical constructions beyond showing that they matched observation, and does not discuss the methods according to which the Neapolitan scholar developed his approach.

Two recent monographs, one by Vincent Ilardi on the history of spectacles and the other one on the history of optics by Olivier Darrigol, mention Della Porta only very briefly.[25] Ilardi gives Della Porta credit for having been the first author to describe complex reflective and refracting systems, but follows Ronchi in judging his theoretical analyses "inadequate".[26] Darrigol, on the other hand, offers a more positive judgment, stating that Della Porta "anticipated the notion of real image", but still devotes less than one page to his contribution.[27] In his recent book on pre-modern and early modern optics, A. Mark Smith presents a detailed assessment of Della Porta's rule for refraction in a glass sphere, noting that it was not meant to be a general theory of refraction and should not be judged as such.[28] As to its empirical adequacy, Smith concludes that it is difficult to establish a posteriori how precise the rule actually was, because of the lack of information on the optical devices employed. Finally, the editors of this volume, Yaakov Zik, Giora Hon and Arianna Borrelli have published detailed analyses of some of Della Porta's optical writings, arguing

[21] Duprè 2000, 2003, 2005, 2006. On Duprè's reference to Lindberg, see Borrelli 2014, 56, 58.

[22] Saito 2008, 35.

[23] Saito 2011, 219–224.

[24] "possibilidade de considerar um instrumento como mediator entre experimentador e objeto de experimentação." (Saito 2011, 280).

[25] Ilardi 2007, Darrigol 2011.

[26] Ilardi 2007, 45–46, 214–217.

[27] Darrigol 2012, 25.

[28] Smith 2015, 344–349.

that his methodology, while certainly driven by an interest in fitting phenomena rather than in stating general principles, nonetheless represented an important basis for the later development of a theory of lenses.[29] In October 2014, together with Friedrich Steinle and Cesare Pastorino, the editors organized a conference at the Technische Universität Berlin, inviting international authorities on the subject to present new research results on Della Porta, his optical work and its early modern context. The prompt and positive response to the invitations showed that the time was ripe for an in-depth discussion on and reassessment of Della Porta's optical work: the conference resulted in a most productive exchange of views whose results are collected in the present volume.

Approaching the Neapolitan scholar's optics with a view to reassessing both its contents and its role in the history of early modern optics requires the combination of different methodologies, all of which are represented in this volume. In the first place, it is necessary to take a closer look at Della Porta's extant published and unpublished writings, becoming aware of the complexity of the task he was undertaking as the first scholar of his time to attempt a systematic exploration of the functioning of glass sphere and lenses. Grasping the specific goals which Della Porta set himself in a given text or experiment is paramount for understanding his methodology. This understanding offers coherent criteria for assessing whether or not Della Porta was "successful" with respect to the task he had assigned himself. In this spirit, three of the contributions in the volume develop detailed analyses of how Della Porta framed and tentatively solved specific optical questions. The contributions deal with themes from Della Porta's three main optical texts: Book XVII of the "Natural magic" from 1589 (Yaakov Zik and Giora Hon), the treatise "On refraction" (A. Mark Smith) and the manuscript draft of the treatise "On the telescope" (Arianna Borrelli). Zik and Hon reconstruct the way in which the Neapolitan scholar approached the question of defining and determining the ambiguous locus referred to in early modern optics as the "point of inversion" ("punctus inversionis"). On the basis of text analysis and schematic empirical reconstructions of the experimental set-ups, Zik and Hon show how the connection between the geometrical focus of a lens and the visual phenomenon of image inversion is much less straightforward than might appear when looking at it from the standpoint of today's geometrical optics and lens theory. The visual point of inversion, Zik and Hon argue, cannot be associated with a specific point-like location in space, since it manifests itself at different apparent positions depending on the relative movement of object, eye and lens. The authors argue that Della Porta was the first scholar of his time to explicitly formulate and tentatively solve this difficulty, and his solution provided a fundamental stepping stone for the research of later authors. A. Mark Smith takes up the parts of "On refraction" in which Della Porta offers his physical explanation and diagrammatic reconstruction of vision, i.e., of the process through which rays of light enter the eye and manage to convey an image of the objects they stem from. Smith highlights the special features of Della Porta's approach by comparing it to those of Alhazen and Kepler and, on the basis of Della Porta's rather

[29] Zik and Hon 2012, Borrelli 2014.

scanty theoretical remarks, offers a tentative reconstruction of the reasoning behind his diagrammatic constructions and optical conclusions. Smith underscores that the Neapolitan scholar's approach to optics was both original and problematic, and shows how Della Porta at times reached profound insights in the mechanism of vision, such as the role of the pupil, but also on various occasions seemingly flouted those same basic principles of geometrical optics which he had clearly stated a few pages earlier. Smith's analysis provides very valuable insights into Della Porta's ambiguous relationship with the Perspectivist theory, of which he rejected key elements, while retaining or adapting other ones in his work. Arianna Borrelli discusses the three extant versions of chapter three of the manuscript draft "On the telescope", which provide evidence of Della Porta's continued interest in optical experiments in the last years of his life. The chapter analyzed is devoted to discussing biconcave lenses, and Borrelli offers a reconstruction of Della Porta's methodology. As already noted by Ronchi, Della Porta did not aim at formulating universal principles of light propagation, but rather at finding out the procedures capable of reproducing specific optical phenomena, such as image inversion, magnification, or images "hanging in the air". According to Borrelli, Della Porta made a radically innovative use of the traditional tools of the Perspectivist optics: optical diagrams and their components, in particular the cathetus line. In Della Porta's hands, optical diagrams became "paper tools" in the sense introduced by Ursula Klein in her work on the history of chemistry.[30]

Beside a careful investigation of Della Porta's writings, a reassessment of his optical work also requires its contextualization in the broadest possible sense. No early modern scholar was a "scientist" in modern sense, but Della Porta's role in the scholarly and courtly milieu of his time is particularly difficult to grasp. The contribution by William Eamon takes up this challenge and draws a broad, impressive picture of the South Italian and European culture to which Della Porta belonged, showing how in those times the investigation of the workings of nature was a deeply political matter, and sketching the courtly and philosophical space in which Della Porta's experiments with mirrors and lenses were situated. Complementing Eamon's paper, which focuses mainly on Southern Europe, the contribution by Fokko Jan Dijksterhuis discusses the way in which the natural magical approach to the investigation of nature was practiced in Northern Europe, and in particular in the Low Countries, and how Della Porta's optical work was received in that context. Dijksterhuis takes as starting point for his investigation the question: which devices could count as "optical instruments" in the early modern context of Northern Europe? He approaches this issue by understanding optics in the broad sense of "controlling and manipulating light, sight and perception" and offers an insightful analysis spanning from the work of the "magus" Cornelis Drebbel to that of Christian Huygens. Passing through a number of known and less known scholars from Northern Europe, Dijksterhuis shows how the telescope belonged to a broad category of natural magical instruments for which no clear-cut boundaries between optics, chymistry and mathematics obtained.

---

[30] Klein 2003.

Finally, the contextualization and reassessment of Della Portas's optical work also presupposes comparing his practices to those of contemporary authors, and this is what the contributions by Riccardo Bellé and Albrecht Heeffer endeavor to do. Bellé offers a comparative study of the way in which the investigation of the law of refraction was pursued by Francesco Maurolico on the one hand, and by Della Porta on the other. The two authors could hardly be more different in their epistemic aims and research methods, but inevitably they both made use of the same tool: the optical diagram. While Della Porta's diagrammatic constructions were often semi-qualitative demonstrations of experimental observations, Maurolico employed them as a means of geometrical proof for his proposed universal law of refraction, although at times he obtained mathematically correct results which contradicted experiment. Optical diagrams and their epistemic function are also at the focus of Heeffer's paper. Heeffer analyses the way in which one specific diagram, which had appeared in Witelo's "Perspective", was taken up first by Della Porta, and later by Kepler and Descartes. He shows how the diagram could serve different epistemic functions, and convincingly argues that, for Kepler and Descartes, it became a powerful heuristic tool which framed in geometrical terms questions about a possible universal law of refraction. In the case of Descartes, according to Heeffer, this approach also showed the path to the modern answer to the question, how a universal law of refraction might look like. In concluding the volume, Zik and Hon offer a look back at the contributions and a few final remarks. They focus on the role of Della Porta as a transition figure from the Perspectivist views on light and vision to the theories of reflection and refraction in optical systems such as lenses and the telescope, which were developed in the course of the seventeenth century and are usually associated with the names of Kepler, Descartes and Harriot. These theories had a more limited scope than the general views of the Perspectivists and formed the basis for modern geometrical optics and its fundamental distinction between "real" and "virtual" images. Selecting and connecting results which emerged in this collection of papers, Zik and Hon endeavor to reassess the role of Della Porta in this crucial transitional period in the history of optics. They do this by reconstructing on the one hand Della Porta's complex relationship with the Perspectivist tradition and, on the other, the ways in which his work represented a necessary, yet often neglected stepping stone for later authors both in methodology and in content.

Thanks to the variety in their contents and methodology, the contributions in this volume complement each other in a most fruitful way, showing how research of Della Porta's optics may lead to fundamental insights not only as far as his work is concerned, but also regarding early modern optics and the culture of natural magic. The contributions show how much work remains to be done and the editors hope that the volume will lead to further research on the subject.

Acknowledgements The editors wish to express their gratitude to Friedrich Steinle and Cesare Pastorino who, together with us, organized the conference which constituted the point of departure for this volume. They contributed to the success of the conference with their suggestions and participation. Specifically, we thank Friedrich Steinle for making possible the conference by accepting to host and financially support it, and Cesare Pastorino for his engagement in organizing the conference program. We also warmly thank all participants of the conference for their presentation and contributions to the discussion. For efficient help with practical issues we also wish to thank Nina Krampitz. Arianna Borrelli acknowledges the support received from the DFG through the projects BO 4062/2-1 (TU Berlin, "Early particle physics") and KFOR 1927 (Institude for Advances Studies on Media Cultures of Computer Simulation, Leuphana Universität Lüneburg). Giora Hon and Yaakov Zik are grateful to the support of the Israel Science Foundation (ISF Grant No. 67/09); it was in this framework that their research on Della Porta was conceived and developed.

References

Borrelli, Arianna. 2014. Thinking with optical objects: glass spheres, lenses and refraction in Giovan Battista Della Porta's optical writings. *Journal of Early Modern Studies* 3: 39–61.

Caverni, Raffaello. 1891. *Storia del metodo sperimentale in Italia*. Vol. 1. Florence: Civelli.

Darrigol, Olivier. 2012. *A history of optics: From Greek antiquity to the nineteenth century*. Oxford: Oxford University Press.

Della Porta, Giambattista. 1558. *Magiae naturalis sive de miraculis rerum naturalium libri IIII*. Naples: Cancer.

———. 1589. *Magiae naturalis sive de miraculis rerum naturalium libri XX*. Naples: Salviano.

———. 1593. *De refractione optices parte libri novem*. Naples: Carlino&Pace.

———. 1962 [1610–1615]. *De telescopio*, ed. Maria Amalia Naldoni. Florence: Olschki.

———. [1610–1615]. *De telescopio* [manuscript]. Biblioteca dell'Accademia Nazionale dei Lincei e Corsiniana, Roma, Archivio Linceo 14. Online version: http://bibdig.museogalileo.it/Teca/Viewer?an=000000935209.

Dupré, Sven. 2000. Mathematical instruments and the 'Theory of the concave spherical mirror': Galileo's optics beyond art and science. *Nuncius* 15: 551–588.

———. 2003. The dioptrics of refractive dials in the sixteenth century. *Nuncius* 18: 1–29.

———. 2005. Ausonio's mirrors and Galileo's lenses: the telescope and sixteenth-century practical optical knowledge. *Galilaeana* 2: 145–180.

———. 2006. Visualisation in Renaissance optics: The function of geometrical diagrams and pictures in the transmission of practical knowledge. In *Transmitting knowledge. Words, images, and instruments in Early Modern Europe*, ed. Sachiko Kusukawa and Ian Maclean, 11–39. Oxford: Oxford University Press.

———. 2008. Inside the camera obscura: Kepler's experiments and theory of optical imagery. *Early Science and Medicine* 13: 219–244.

Goulding, Robert. 2013. Thomas Harriot's optics, between experiment and imagination: The case of Mr Bulkeley's glass. *Archive for History of Exact Sciences* 68: 137–178.

Hoppe, Edmund. 1926. *Geschichte der Optik*. Leipzig: Weber.

Ilardi, Vincent. 2007. *Renaissance vision from spectacles to telescopes*. Philadelphia: American Philosophical Society.

Klein, Ursula. 2003. *Experiments, models, paper tools. Cultures of organic chemistry in the nineteenth century*. Stanford: Stanford University Press.

Kodera, Sergius. 2015. Giambattista della Porta. In *The Stanford encyclopedia of philosophy*, ed. N. Edward. Zalta http://plato.stanford.edu/archives/sum2015/entries/della-porta/.

Lindberg, David. 1976. *Theories of vision from al-Kindi to Kepler*. Chicago: University of Chicago Press.

———. 1984. Optics in sixteenth-century Italy. In *Novità celesti e crisi del sapere*, ed. Paolo Galuzzi, 131–148. Florence: Giunti.

Lohne, Johannes. 1963. Zur Geschichte des Brechungsgesetzes. *Sudhoffs Archiv für Geschichte der Medizin und der Naturwissenschaften* 47: 152–172.

Naldoni, Maria Amalia. 1946. L'ottica del Porta dal 'De refractione' al 'De telescopio'. *Atti della Fondazione Giorgio Ronchi* 1: 54–72.

———. 1962. Un manoscritto inedito di G.B. Della Porta. In *De telescopio*, ed. Della Porta, Giovan Battista, and Maria Amalia Naldoni, 21–28. Florence: Olschki.

Priestley, Joseph. 1772. *History and present state of discoveries relating to vision, light and colour*. London: Johnson.

Rienstra, M. Howard. 1975. Giambattista Della Porta. *Dictionary of Scientific Biography* 11: 95–98.

Ronchi, Vasco. 1947. Storia delle lenti. *Atti della Fondazione "Giorgio Ronchi"* 2: 1–24.

———. 1962. Du "De refractione" au "De telescopio", In *De telescopio*, ed. Della Porta and Giovan Battista, Maria Amalia Naldoni, 1–19. Florence: Olschki.

———. 1970. *The nature of light. A historical survey*. London: Heinemann (original Italian editions 1939).

Saito, Fumikazo. 2008. Experimento e matemática: o estudo das lentes segundo a perspectiva de Giambattista della Porta (1535–1615). *Circumscribere* 4: 83–101.

———. 2011. *O telescópio na magia natural de Giambattista della Porta*. São Paulo: EDUC/Livraria da Fisica Editorial.

Smith, A. Mark. 2015. *From sight to light: The passage from ancient to modern optics*. Chicago: University of Chicago Press.

Wilde, Emil. 1838. *Geschichte der Optik. Erster Teil*. Berlin: Rücker und Püchler.

Zik, Yaakov, and Giora Hon. 2012. Magnification: How to turn a spyglass into an astronomical telescope. *Archive for History of Exact Sciences* 66: 439–446.

Chapter 2
A Theater of Experiments: Giambattista Della Porta and the Scientific Culture of Renaissance Naples

William Eamon

Abstract Renaissance Neapolitan scientific culture developed within the framework of its academies, in the shadow of the Spanish empire, and under the philosophical spell of Bernardino Telesio and Giambattista Della Porta. The academies occupied a prominent place in the landscape of Neapolitan culture. The Neapolitan scene is distinguished by the predominance of salons devoted to theatre and the search for secrets of nature. The academies were often covers for discussion about the political realities of Naples, in particular its status as a subject territory of the Spanish Monarchy. Telesio's philosophical naturalism framed the southern Italian philosophical tradition, which came to its fullest expression in the ideas of Della Porta and Tommaso Campanella. Following Giambattista Della Porta through his life and influence, this essay attempts to provide a deep description of scientific culture in Renaissance Naples

When the famous Bolognese surgeon and professor of secrets Leonardo Fioravanti lived in Naples in the 1550s, a wizard showed him a magic mirror that terrified and delighted him. "The mirror looked like any other," the surgeon wrote, recalling the event years later, "but was so artfully made that when a person stood before it to view himself he saw more than a dozen figures, like phantoms, come out of it. It struck terror in all who set eyes on it. Nothing more astonishing (*mostruosa*) in this art had ever been seen before."[1]

Who might have shown Leonardo the marvelous magical mirror that made "monstrous" images appear in it? Fioravanti didn't identify the wizard of Naples, but a likely candidate was the precocious Neapolitan magus Giambattista Della Porta. Fioravanti identifies the wizard as an aristocrat, "a gentleman"—and so Della Porta was, the strapping young son of a wealthy Neapolitan nobleman. Like Fioravanti, he would soon become a celebrity, first in the ornate cultural world of Naples, then all of Italy, and at length Europe itself. A single book made Della Porta

[1] Fioravanti 1567, 55.

W. Eamon (✉)
New Mexico State University, Las Cruces, NM, USA
e-mail: weamon@nmsu.edu

A. Borrelli et al. (eds.), *The Optics of Giambattista Della Porta (ca. 1535–1615): A Reassessment*, Archimedes 44, DOI 10.1007/978-3-319-50215-1_2

Fig. 2.1 View of Naples as it appeared in the late fifteenth century; completed around 1473 by Francesco Pagano. Tavola Strozzi, Museo Nazionale di San Martino (Photoserve Electra)

a superstar: his phenomenally successful *Magia naturalis*, in which the great magician depicted lenses and mirrors that are identical to the one that so captivated Fioravanti. One of them, Della Porta wrote, "makes multiple representations, so that many things are seen at once; opening and shutting it, you will see twenty figures for one, and more." Della Porta's famous experimental demonstrations made his residence, the Villa Pradelle in nearby Vico Equense, a coveted stop on the European virtuosi's Grand Tours of Italy. For visitors lucky enough to have an introduction from a prominent intellectual or writer, Della Porta would do tricks with loadstones to demonstrate their surprising properties, and display optical devices such as mirrors that projected "diverse images one after another, and of diverse color," mirrors casting reflections of reflections of themselves, and one that showed images of the viewer suspended in the air, "like birds in flight."[2] All were the awesome productions of the new Renaissance science of natural magic.

Della Porta would within a few decades become Renaissance Europe's most celebrated magus, and Fioravanti Italy's most famous professor of secrets—although when the two met Della Porta was a mere teenager and Fioravanti was still an unknown surgeon. Della Porta was just beginning his scientific research. In doing his experiments, he had organized an academy, the *Accademia dei Secreti*, that met periodically in the family palace near the Via Toledo in the center of Naples.[3] There the young aristocrat and his companions busied themselves with experiments that would demonstrate the power of natural magic as a way of grasping nature's secrets, while entertaining visitors like awe-struck Leonardo Fioravanti.

Della Porta and his companions were not alone in this enterprise. Contemporaries dubbed experimenters like Della Porta and Fioravanti the "professors of secrets." The brilliant and eccentric social critic Fra Tomaso Garzoni created an indelible portrait of them in his *Piazza universale di tutte le professione del mondo* (Universal Plaza of All the World's Professions, 1580). Garzoni depicted them as a community of hyper-charged experimenters who burned with such zeal for discovering "secrets of nature" that they "desired them more than life's necessities."[4] Lorraine Daston and Katharine Park, in *Wonders and the Order of Nature*, describe them as "preternatural philosophers" who explored "the large and nebulous domain of the marvelous" that stood midway between the commonplace and the miraculous.[5] It was a domain that had long resisted philosophical or causal explanation. The Scholastic philosophers never disputed the existence of anomalous events that popped up unexpectedly in the course of nature. The world was full of them: the lodestone's mysterious attraction of iron, the virtues of bezoar stone, and so forth. The famous *Book of Secrets* attributed to Albertus Magnus described dozens of them.[6] With few exceptions, however, medieval intellectuals excluded such phenomena from the

---

[2] Della Porta 1957, 356–7.

[3] Gliozzi 1950.

[4] Garzoni 1996, 324. On Garzoni's *Piazza universale*, see Martin 1996.

[5] Daston and Park 1998, 159.

[6] Pseudo-Albertus 1973.

purview of natural philosophy. By contrast, the Secreti's experiments were all designed to produce—and thereby discover—preternatural phenomena.

Many of Della Porta's experiments had practical applications, including dyeing fabrics, distilling cosmetic waters, and making steel, while others, such as using natural magic to sire multicolored colts, strained credibility.[7] Whether practical or illusory, the experiments were *performances* designed to provoke wonder and amazement. In fact, Della Porta reports, no one was admitted into his academy unless he had "brought forth some admirable new secret of nature beyond ordinary comprehension."[8] Della Porta was quite a prodigy. He claimed that he wrote his blockbuster *Magia naturalis* at the tender age of fifteen, though the book wasn't published until 1558, when he was barely 23.[9] If Della Porta was Fioravanti's mysterious gentleman, the precocious magus would have been a teenager when the two met.[10]

We know nothing of the circumstances under which the surgeon and the magus met. It might have been in Della Porta's academy, where the youthful magician would have demonstrated the mirror that so fascinated Fioravanti—provided the surgeon had brought his admission ticket, a hitherto unknown "secret of nature." More likely, the two met in another academy, coincidentally also called the *Accademia Segreta*—as far as we know the first of several academies of secrets to be formed in and around Naples in the 1540s. Girolamo Ruscelli, one of the group's members, reported that the academy pledged "to make diligent inquiries and, as it were, a true anatomy of the operations of nature in and of itself"[11]

The Segreti proposed to make "true anatomies" of nature by means of experiments—trying out medical and craft recipes, concocting all manner of distilled products, and making new alloys and precious metals. Of course they didn't do the experiments themselves, because they were gentlemen and scholars: they hired gardeners to grow the herbs they needed for their trials and craftsmen to operate the furnaces and alchemical apparatus in their laboratory, which they called the "Filosofia." The helter-skelter, chemistry-kit style of experimenting that Ruscelli describes was typical of the professors of secrets.

The Academy of Secrets published its discoveries in a bestselling book of secrets, *I Segreti del reverendo dono Alessio Piemontese*, which came out in 1555, 3 years before Della Porta's *Magia naturalis*. The book's author, "Alessio Piemontese" (a pseudonym of the humanist and popular writer Girolamo Ruscelli) prefaced the work with the heartbreaking story of a reclusive monk who revealed his secrets only after learning that he'd let another doctor's patient die rather than reveal a secret that

---

[7] Eamon 1994.

[8] Della Porta 1957, 2.

[9] Della Porta described mirrors similar to the one that Fioravanti mentions in *Natural Magic*, 356, 357.

[10] Fioravanti lived in Naples from 1550–1555, when Della Porta would have been between fifteen and twenty years old.

[11] Eamon and Paheau 1984. The Segreti, probably sponsored by the powerful Neapolitan nobleman Alfonso d'Avalos, Duke of Salerno, may have met in both Naples and Salerno. Ruscelli was at one time employed in Avalos's court.

might have cured him. Grieving over his stinginess, Alessio vowed to publish all his secrets. The sorrowful tale of Alessio Piemontese would reverberate throughout early modern Europe, inspiring generations of professors of secrets to publish their experiments. They saw how famous the strategy made Alessio.[12]

The Academy of Secrets that Ruscelli described was in all likelihood the inspiration for Della Porta's academy. The similarities between the two are unmistakable. Leonardo Fioravanti also emulated the Segreta in organizing an academy of his own in his Neapolitan lodgings. Not surprisingly, the three groups overlapped in membership and were closely interconnected with other academies. So there were numerous opportunities for Della Porta, Ruscelli, and Fioravanti to have met one another in Naples, a city teeming with informal academies and salons. The magician, the humanist, and the surgeon surely crossed paths.

To the modern reader Fioravanti's description of awestruck spectators gawking at Della Porta's magic mirror seems a little quaint. So accustomed are we to similar illusions—and myriads of others far more electrifying and convincing, in film, on stage, and in real life—that the notion a simple distorting mirror might "strike terror" into spectators seems, to say the least, a little remote.

Remoteness can be to the historian's advantage, however. Viewing the past from an alienating distance can provide a wider perspective on the scientific culture of Renaissance Naples. In addition to seeing Della Porta's demonstrations as scientific experimentation in optics and physics, we can also observe them in the context of performance and audience—because Della Porta's experiments were, above all, *demonstrations*. Only they weren't demonstrations of the *theoretical principles* of natural magic, but of the inherent power of occult forces and the magus's ingenuity and skill in manipulating them. Manifestations of marvelous but natural forces, they are preternatural events choreographed by the magus.[13] Like John Tyndall's famous scientific demonstrations three centuries later, or Jean-Antoine Nollet's reported demonstration of an electrostatic machine that shocked 200 Cistercian monks, they were performed onstage with an enthralling technical setup, visually displaying the costume and paraphernalia of the magus-scientist.[14] In the theater of magic—whether of natural magic or experimental science—the magus is both director and actor in the drama that unfolds.

As Louise Clubb first pointed out, and Sergius Kodera further elaborated, Della Porta achieved the same magical effects in his theatrical writings, with their convoluted plots, exaggerated effects, and *meraviglie*.[15] Lovers are never merely in love in Della Porta's plays; they are always besotted. A beautiful woman is never merely pretty, but instead like Clarizia in *The Two Rival Brothers*, marvelous, almost supernatural. "She was so beautiful," Della Porta writes, "that you couldn't tell whether beauty made her beautiful or she made beautiful beauty itself So dazzling

[12] Ruscelli revealed that he was the author of "Alessio's" *Secreti* and disclosed the information about the Accademia Segreta in his *Secreti nuovi* (Venice 1567). See Eamon and Paheau 1984.

[13] Kodera 2014.

[14] De Young 2011; Riskin 2008, 58.

[15] Clubb 1965; Kodera 2012, 2014.

a spectacle of beauty ravished to itself the eyes and hearts of all observers: tongues fell silent, spirits hung suspended."[16] Like Della Porta's magic mirror, she made everyone "all eyes" when gazing at her. Clarizia imparts an otherworldly sense of beauty, like a miracle.[17] The *meraviglie* that Della Porta produced in his theater of magic were staged to appear equally miraculous.

This is courtly science; it presages baroque science. The magus's quest for secrets, his craving for rarities, his incitement to wonder, and the notion of science as a theatrical performance aiming to delight and amaze perfectly fit the baroque, courtly manner. In *Malpiglio*, Tasso's dialogue on the court, one of the interlocutors observes that Naples is a city "where a man's worth shines out more clearly in courts than anywhere else."[18] Reading Della Porta's *Magia naturalis* is to plunge headlong into the cultural world of the Renaissance court. His "catalogue of rarities" contained nothing common or base, he assured, just exquisite, curious, and ingenious things fit for a gentleman. Knowing secrets separated noble minds from the vulgar, who were unworthy of the art.

Dissimulation was a necessary talent in the performance of secrets. The wizard has to keep secrets from falling into the hands of ignorant people, making a show of it by concealing the trick, since ignorance of the trick enhances wonder. As practically every treatise on courtly behavior pointed out, dissimulation—commonly called prudence—was also necessary to survive in the world of the courts. Natural magic was not merely entertainment; it was also a strategy to accommodate to the political realities of Renaissance Naples.

The central political reality of Naples in Della Porta's time was its status as a subject territory of the Spanish Monarchy. Ruled by a Spanish Viceroy and host to hundreds of Spanish noblemen, soldiers, literati, and fortune seekers, Spanish Naples was in every way a royal city—except for the absence of a king. The residence of 13 princes, 24 dukes, 30 marquises, and countless lesser nobility, Naples was "the jewel of the Spanish Crown."[19]

Neapolitan aristocrats remembered the early Spanish viceroys as benevolent rulers who treated citizens with respect and paternal love. Things changed with the arrival in 1532 of Don Pedro of Toledo whose job was to institute the imperial policy of the Spanish Monarchy.[20] To the local barons, with the arrival of Toledo the viceroys began behaving more like supreme masters who no longer recognized the aristocracy's privileges than as the compassionate rulers of the past.

When Toledo arrived in Naples to assume his duties, he found a kingdom wracked by decades of baronial infighting, foreign invasion, and unrelenting pestilence. He quickly enacted measures to restore public order and bring the unruly barons to heel.[21] These measures angered the nobles and conflict between the Viceroy and the

---

[16] Della Porta 1980, 59.

[17] Della Porta 1980, 39.

[18] Tasso 1982, 181.

[19] Marino 2011, 27.

[20] Musi 2007.

[21] Toledo's reign as Viceroy is treated in a comprehensively in Hernando Sánchez 1994.

nobility intensified. Toledo's attempt to introduce an Inquisition based on the Spanish model, which made the tribunal a state agency, gave barons the pretext they needed for open rebellion. In May 1547, when the edict announcing the Inquisition was posted, the city exploded in revolt.[22]

Although the Spanish Inquisition is often regarded as an implacable, retrograde institution staffed by zealots who brutally defended Spain's religious purity, it was in fact one of the most modern bureaucratic institutions of its time.[23] Far from being a fanatical instrument for imposing religious orthodoxy, the Spanish Inquisition was, as Henry Kamen pointed out, "in every way an instrument of royal policy, and remained politically subject to the crown."[24] Its main task was to combat perceived threats to undermine the Spanish state, whether by heretics, Judaizers, or practitioners of magic. Like the Royal Protomedicato, Spain's health ministry, its reach extended throughout the empire. The Inquisition was a worldwide network designed to advance the interests of the Monarchy in Spain and in its colonies. In that sense, one might say that the Spanish Inquisition was a vehicle of modernity.[25]

The academies occupied a prominent place in the landscape of Neapolitan culture.[26] About fifty academies existed in Naples during the sixteenth and early seventeenth centuries.[27] Although philosophical and literary academies sprang up in many sixteenth century Italian cities, the Neapolitan scene stands out not only for the number of its academies but also by the predominance of groups devoted to theatrical productions and the hunt for "secrets of nature." Spurred on by the patronage of Ferrante Sanseverino, the Prince of Salerno, wealthy Neapolitans delighted in dressing up and reciting comedies for each other in private salons.[28] Della Porta wrote many of his plays for production at these amateur gatherings. At least half a dozen academies devoted to experiments and "secrets" can be identified in the mid-sixteenth century Naples, including a group calling itself the Incogniti, which delved into the "ten books on the mysteries of nature" composed by "Segreto," one of its members.

The Neapolitan academies were not just artistic and philosophical gatherings, however. They were also places of aristocratic self-fashioning: they cultivated *conoscenza di sè*, or "self-knowledge." Discussions about self-knowledge led inevitably to deliberations about the status of the aristocracy, whose privileges were under siege. Beneath the façade of amateur theater and playing with mirrors, members of the academies toyed with a dangerous political agenda. It was one thing to sponsor amateur theatrical troupes; it was quite another to participate in clandestine gatherings that harbored anti-Spanish political sentiments. As the Viceroy sus-

[22] Eamon 1994, 154.

[23] Romeo 2013.

[24] Kamen 1997, 137.

[25] Silverblatt 2004.

[26] Canepa 2013.

[27] Everson 2012.

[28] Clubb 1965, 8–9; Wistreich 2007, 24.

pected, the academies were seedbeds of revolution, and when the revolt broke out in the summer of 1547, Toledo brutally suppressed them.[29]

We don't have to speculate that the academies bred talk of revolution. Contemporaries attest to it. Antonio Castaldo, a member of one of the academies, explained:

> The reason [why Toledo disbanded the academies] was that it did not seem a good thing that under the pretext of literary activities, there were so many assemblies and virtually of the brightest and most important minds in the city, both nobles and ordinary citizens, since through study of the humanities, people made themselves confident and informed, and also became more spirited and resolute in their actions.[30]

Writing in the 1580s, when memory of the revolt was still alive, the historian Scipione Mazella stated it more bluntly: "The Viceroy became suspicious that the academies were fostering groups that denounced the king."[31] In other words, the academies were considered cover for political debate and organization.

Castaldo didn't divulge the specifics of the discussions that took place in the academies. But it's possible to reconstruct some of what transpired. We know that magic, astrology and alchemy were all topics of discussion. Toledo was deeply troubled by the Neapolitan intellectuals' curiosity about such matters. In 1544, when the subject of alchemy came up in conversation at his court, the viceroy asked his advisor, the humanist Benedetto Varchi, his opinion about the art. Varchi warned the Viceroy that while some forms of alchemy are legitimate, "sophistical" alchemy—the kind that claims to transmute gold or make marvelous panaceas—"should be persecuted with fire."[32]

In investigating nature "in and of itself" (as Ruscelli put it), Neapolitan intellectuals found common ground in the naturalist philosophical tradition that descended from the teachings of Bernardino Telesio (1509–1588), a charismatic Dominican friar from Cosenza.[33] Telesio, whom Francis Bacon dubbed "the first of the moderns," waged a frontal assault on Aristotelian metaphysics. He taught that to get at nature's true character, it must be investigated "according to its own principles" (*iuxta propria principia*) and be allowed to "speak for itself." Rejecting Aristotelian metaphysics, Telesio reduced the principles of his system to the dueling forces of heat and cold, which battle to gain possession of passive matter. Out of that perpetual Heraclitean struggle arise the qualities of things.[34]

If to Telesio nature was concrete and physical, it was not inert, like the mute atoms of Lucretius. Things did not arise by blind chance. Telesio's nature was alive, sentient, and pulsing. Endowed with sensation, nature "feels" the shapes and forms appropriate to it. Radically anti-Aristotelian and uncompromisingly materialistic,

---

[29] Eamon 1994, 154.

[30] Minieri-Riccio 1879, 173.

[31] Hernando Sánchez 1994, 500.

[32] Varchi 1827, 25.

[33] Cocco 2013a.

[34] Van Deusen 1935.

Telesio's naturalism was uplifting, liberating, and dangerous, for it implied that you could subject social and religious norms to the same empirical scrutiny with which you looked at nature, and could use nature as the standard to judge them. In the fecund intellectual soil of Italy's South naturalism embraced a reform agenda, taking aim at social and religious conventions. And by introducing a concept of nature centered on the aliveness things Telesio reopened the problem of how magic works. Telesio's ideas left an indelible mark on the southern Italian philosophical tradition.[35] "In the intellectual and cultural world of Spanish Naples," writes historian John Marino, "the whole reign of Philip II lay under the double spell of Telesio and Della Porta."[36]

Spain's aura was visible everywhere in this colonial city—not only in the presence of soldiers, aristocrats, and government ministers but also in the city's architecture, upon which the Crown lavished extravagant expenditures.[37] The popularity of the picaresque novel, whose trickster heroes seemed to imitate the social ups and downs of court life, was also Spanish. The Spanish influence was visible in the natural sciences too, particularly the Catalonian alchemical tradition that descended from writings attributed to Ramon Lull and Arnald of Villanova, two thirteenth-century Catalonian friars. The pseudonymous alchemical writings attributed to Arnald and Lull nurtured a scientific underworld in Naples that fused alchemy with social and political reform. In contrast to the Paracelsian school that flourished in northern Europe, which elaborated a complex, all-encompassing alchemical philosophy of nature, the Catalonian school made distillation its nearly exclusive domain.[38] To the alchemists of this school, distillation was a technology capable of purifying organic substances from a state of decay, transforming coarse material into clear, transparent liquors that preserved the traits of the original, but in a refined and exalted form.[39] It also served as an explanatory model to account for phenomena that were difficult or impossible to observe, such as digestion and the interaction of soul and body. In private salons, Neapolitan intellectuals conjectured about the symbolic meanings of distillation, seeing the alchemical art as a metaphor of social reform, a purification of society by separation similar to alchemical separation, purging Neapolitan culture of foreign influences and recovering the pure Neapolitan society of olden times. Distillation, the art of separating the pure from the impure, became the rage in the scientific underworld that buzzed in the Neapolitan academies.

Della Porta could barely contain his enthusiasm for alchemy, calling it the "art to be preferred above all others." Because it emulates nature, alchemy displays natural processes in an environment controlled by the magus. "This most useful art," wrote Della Porta, "emulates nature, assisting it in producing wondrous effects beyond

[35] Badaloni 1980; Agrimi 1992. The articles by Badaloni cited in this chapter are also in Badaloni 2004.

[36] Marino 2002, 7.

[37] Cummins 2015.

[38] Pereira 1990.

[39] Moran 2005.

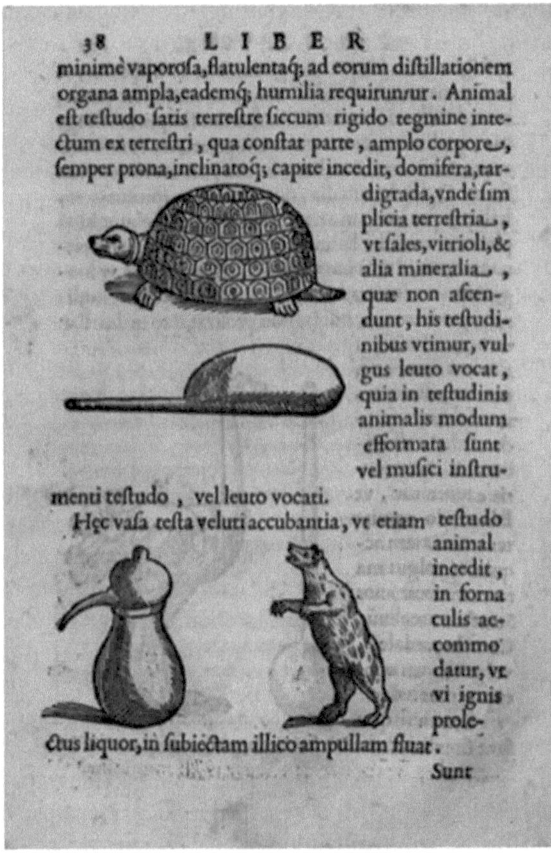

38 LIBER

minimè vaporoſa, flatulentaq; ad eorum diſtillationem organa ampla, eademq; humilia requiruntur. Animal eſt teſtudo ſatis terreſtre ſiccum rigido tegmine intectum ex terreſtri, qua conſtat parte, amplo corpore, ſemper prona, inclinatoq; capite incedit, domifera, tardigrada, vndè ſimplicia terreſtria, vt ſales, vitrioli, & alia mineralia, quæ non aſcendunt, his teſtudinibus vtimur, vulgus leuto vocat, quia in teſtudinis animalis modum efformata ſunt vel muſici inſtrumenti teſtudo, vel leuto vocati.

Hęc vaſa teſta veluti accubantia, vt etiam teſtudo animal incedit, in fornaculis accommodatur, vt vi ignis prolectus liquor, in ſubiectam illico ampullam fluat.

Sunt

Fig. 2.2 Distillation vessels from Giambattista Della Porta, *De distillatione* (1610)

any ever seen or scarcely believed possible." Applying the science of physiognomy, he concluded that alchemical vessels should be designed in "the imitation of nature." Flatulent humors, for example, need to be distilled in bulky vessels, while thin spirits must be drawn through a long, narrow passage to greater heights in order to purify them (Fig. 2.2). As Della Porta lyrically expressed it, the art of distillation "follows and resembles the showers and dew of heaven, as the daughter the mother."[40]

Spain put its stamp on Neapolitan scientific culture in other ways besides nurturing alchemy. Della Porta was a Neapolitan citizen but also a Spanish subject who belonged to a vast imperial system stretching from Italy to the Andes, an empire, in the words Fray Francisco de Ugalde "on which the sun never set" (*el imperio en el que nunca se pone el sol*). The burden of outside control rested uneasily on Spain's colonial subjects, causing them to turn their gaze toward local difference and singularity. Discovering the landscape was a means of expressing identity.[41] In their pur-

[40] Della Porta 1957, 254–5.

[41] Cocco 2013b, 9.

suit of natural history, Neapolitan intellectuals were no different than Spanish subjects elsewhere. As Jorge Cañizares explains, "Members of the local intelligentsia of every kingdom that constituted the Spanish Empire, from Naples, Sicily, and Aragon and Mexico were heavily invested in developing patriotic surveys of local material and spiritual resources."[42] Doing natural history was a patriotic act and a means to define identity.

Neapolitans inherited a rich natural history tradition going all the way back to the first-century Roman naturalist Pliny the Elder, who succumbed to noxious fumes from an eruption of Mount Vesuvius while observing from close by. Volcanologists still classify major volcanic eruptions as "Plinian." Ferrante Imperato (1550–1625), a polymathic apothecary, collector, naturalist, and acquaintance of Della Porta emulated Pliny's encyclopedic curiosity in his book on natural history, *Historia naturale*, published in 1599. Rocks and minerals made up the bulk of Imperato's ambitious work—a reflection of his personal collection of fossils and other naturalia on display in his museum in the Palazzo Gravina in central Naples. Imperato was an indefatigable collector of naturalia. His cabinet of curiosities, a menagerie of objects ranging from shells and stuffed animals to petrified objects and an alligator hanging from the ceiling, made his residence a prime destination for European intellectuals visiting Naples. Paula Findlen observed that the mania for collecting objects of natural history merged humanist scholarship with empirical, fact-based inquiry and blurred the boundary between the ancients and moderns.[43] Visuality was a predominant attribute of Neapolitan scientific culture. Whether displaying nature in collections such as the one in Imperato's museum or performing nature in Della Porta's Villa delle Pradelle outside Naples, where the magus entertained guests, Neapolitan intellectuals knew nature by visualizing it (Fig. 2.3).

Imperato's museum and Della Porta's villa were both spaces for displaying nature and performing experiments. But they had completely different purposes. Paula Findlen explains that the two spaces typified entirely different trends within the Renaissance naturalist community. Della Porta saw nature as infinitely metaphorical and full of mysterious and meaningful resemblances. His villa was a performance space, in which experiments were spectacles that provoked wonder. His collection was imbedded in the humanist tradition and was amply furnished with objects of classical antiquity, including Greek and Roman statues, medals, books, and manuscripts. That's because Della Porta always wanted to *read* nature and understand its relation to the body of magical knowledge. Imperato's museum, on the other hand, was arranged as a scientific instrument for inspecting and cataloging nature. "Imperato professed to study nature *only from nature* and for the betterment of medicine, while Della Porta perceived natural history as a prelude to natural magic."[44] Unsurprisingly, Imperato was Della Porta's main rival as the the preeminent naturalist in Naples.

---

[42] Cañizares-Esguerra 2006, 9.

[43] Findlen 1994, 316–7.

[44] Findlen 1994, 227.

Fig. 2.3 Ferrante Imperato's museum in the Palazzo Gravina, Naples. Engraving from Ferrante Imperato, *Dell'historia naturale* (Naples 1599)

Landscape influenced Neapolitan scientific culture in myriad ways. A proverb had it that *"Napoli è un paradiso abitato da diavoli"* ("Naples is a paradise inhabited by devils). Coined during Angevin times, the proverb counter poses the grandeur of the volcanic landscape and the richness of the Campania against the squalidness of urban life in the crammed city.[45] The extremes that so marked the economic and political life of Naples were not lost on its philosophers, either. And no one was more awed by the terror and wonder of the Neapolitan landscape than Bernardino Telesio, who transposed the miracle of its prodigiousness into metaphysics.

Living in the shadow of Vesuvius, the hulking mountain that everyone knew could at any moment explode with violent force, Neapolitans were constantly in the presence of nature's occult powers. Vesuvius had last rumbled in 1500 in an ominous burp of steam and waited until a few hours before dawn on 16 December 1631, when explosions ripped gigantic cracks in the base of its massive cone. The mountain glowed a terrifying red, then vanished into darkness, covered in billowing clouds of ash.[46] One amazed eyewitness recorded detailed daily observations of the eruption in a work he titled "the stupendous workings and marvelous excesses of nature produced by Mount Vesuvius" (*L'opre stupende e merauigliosi eccessi dalla*

[45] Marino 1982.

[46] Cocco 2013b, 58.

natura prodotti nel Monte Vesuvio della cittá di Napoli, 1632).[47] Pietro Asterio, who survived the eruption, mused that the prodigious event gave occasion to reflect on "the curious labor of discovering the secrets of nature."[48] What caused Vesuvius's eruptions? What is the earth's crust made of? Were the eruptions portents of future calamity? Why is the earth hot within? Such questions stirred animated debate. Natural philosophy, alchemy, and astrology were all marshaled to explain Vesuvius's unpredictable and violent behavior.[49]

To the Neapolitan naturalists, Vesuvius's fiery eruptions confirmed Telesio's speculations about the centrality of heat in the workings of nature. Imperato dedicated a lengthy chapter of his *Historia naturale* to "the origin of fire and the various operations of heat and cold." Even while Vesuvius lay dead cold, Neapolitans witnessed the power of heat in Pozzuoli and the Campi Flegrei. These places, and the icy mountain peaks of the Campania in winter, bore witness to the perpetual struggle between the polar opposites that underlay natural phenomena. Telesio's philosophy was a natural fit for Neapolitans, who witnessed nature's prodigious powers in their daily lives. With its rigorous adherence to sensory experience, Telesio's teachings inspired an entire generation of Neapolitan intellectuals striving to rethink Aristotle's authority by relying strictly upon empirical evidence from the natural world.

Della Porta published his academy's experiments in his book, *Magia naturalis*, which appeared in 1558, just 3 years after the publication of Alessio Piemontese's *Secreti*. He dedicated the work to the young King Philip II of Spain, who had acceded to the throne 2 years earlier. Then Della Porta embarked on a customary Grand Tour of Europe, where he consulted libraries and met natural philosophers in Germany, France, and Spain. His most important stop was Madrid, where the young magus presented his book on natural magic and another on cryptography—sciences that locked and unlocked the secrets of nature—to the young king.[50]

Della Porta would have found Philip's court to be a space perfectly suited to his style as a magus. As Elliott and Brown write, "The court of the King of Spain resembled a magnificent theater in which the principal actor was permanently on stage. The stage instructions were meticulously detailed; the scenery was imposing, and the supporting cast was impressively large."[51] Courtiers in Philip's retinue might have stepped straight from the pages of Castiglione's *Book of the Courtier*, where an interlocutor exclaims, "Look at the Spaniards, who seem to be the masters in courtiership".[52]

What impression Della Porta's magic may have made on Philip is impossible to say. No record of their meeting survives. He certainly wasn't the first Italian with

---

[47] Cocco 2013b, 54–5.

[48] Cocco 2013b, 84.

[49] Everson 2012.

[50] Although we don't know Della Porta's exact itinerary, he probably visited Spain in the mid-1560s. See Clubb 1965, 13.

[51] Brown and Elliott 1980, 31.

[52] Castiglione 1959, 115.

big ideas to seek the king's patronage. The son of a minor Neapolitan nobleman, he couldn't claim an impressive lineage and didn't yet have the reputation he would later gain as a magus. Like most visitors who had an audience with the distant and private king, Della Porta was probably greeted politely, then dismissed with a few anodyne remarks whose exquisite courtesy never failed to impress. But Della Porta, an upwardly mobile aristocrat, knew that the audience with the King of Spain would boost his credentials.

Returning to Naples after his grand tour Della Porta, an aristocrat trained in the courtly art of prudence, accommodated to the political order. His comedies catered to a Spanish audience, gradually becoming more and more "Hispanic" as he matured as a playwright. According to Della Porta biographer Louise Clubb, he "yielded to Spanish influence and to Counter-Reformation demands for increased didacticism".[53] Having friends in high places helped him in 1574, when he was summoned by the Inquisition and questioned about his views on magic and witchcraft. The Tribunal let him off with a slap on the wrist and a warning—which, needless to say, he heeded.[54]

Della Porta's reputation soared. A second, expanded edition of the *Magia naturalis* was published in 1589, displaying his nonstop experimental activity since the first edition. The book was a sensation when it came out in Italian (15 editions) and was translated into English, French, German, Dutch, and Spanish. *Magia naturalis* was for four generations the theoretical foundation for natural magic all over Europe. No European intellectual who kept up with developments in the sciences of secrets could ignore it.[55] No wonder it was said that the two greatest tourist attractions of Naples around the year 1600 were the baths of Pozzuoli and Giambattista Della Porta.[56]

One of the tourists who came calling was the French virtuoso Nicolas-Claude Fabri de Peiresc (1580–1637), who visited Della Porta in 1601 while touring Italy. Peiresc carried letters of introduction to Della Porta and to Imperato, who by this time were the two leading lights of Neapolitan scientific culture. Imperato's museum was as famous among the European literati as Della Porta's villa.[57] They represented two completely different scientific spaces, however. Della Porta received visitors at the family villa as would any aristocrat: with grace and an eye toward entertainment. He liked Peiresc so much that he treated the young virtuoso to a few experiments. Gassendi, Peiresc's close friend and biographer, intimates that they were probably similar to the ones the magus enacted for other visitors: demonstrations of optical illusions, burning mirrors, and "merry conceits of the loadstone," such as the Ouija board with a wooden figurine in a rowboat in a container of water, surrounded by the letters of the alphabet, which, through a clever manipulation of a hidden magnet, mysteriously spelled out words to the amazement of viewers—in other words, the

[53] Clubb 1965, 66.

[54] Clubb 1965, 52; Aquilecchia 1968; Lopez 1974.

[55] For a summary treatment of Della Porta's natural science, see Muraro 1978.

[56] Clubb 1965, xi.

[57] Stendardo 2001.

very demonstrations that filled Della Porta's *Magia naturalis* and that so fascinated Leonardo Fioravanti.[58]

While Della Porta allowed Peiresc to witness the intimate workings of his experimental theater, the French virtuoso's visit to Imperato's museum barely elicited a mention in Gassendi's biography. Though Peiresc judged the pharmacist's cabinet "well furnished with rarities of nature," he concluded that compared to Della Porta, "an eager searcher out and explainer," Imperato was merely "a diligent keeper and preserver of the most abstruse works of nature." Peiresc found Della Porta's science thrilling and steeped in profound mystery and symbol. The conversations between the two virtuosi about the secrets of nature filled many hours. In contrast, Imperato's museum was packed full of things without any of the paraphernalia of erudition that made science humanistic. As Paul Findlen observes, "Imperato collected objects but did not possess 'wisdom,' in the humanist sense of the term, and therefore could not provide the learned conversation that initiated the sharing of secrets".[59] Yet in possessing nature in all its variety and wonder and particularity, Imperato was furthering a different way of scientific inquiry in which particulars, for all their peculiarity, would be the focus of research under the reign of a new factual sensibility that took shape in the late Renaisssance.[60]

In the spring of 1604 Federico Cesi (1585–1630), the 18-year-old marquis of Monticello, visited Della Porta. Cesi was struggling to save from dissolution a recently founded "brotherhood of searchers of the arcane sciences," the Accademia dei Lincei (Academy of Lynxes). Cesi had founded the society as an attempt to put Della Porta's ideal of science as a quest for rare secrets of nature into practice.[61] The Lincei took their emblem from the *impresa* Della Porta had chosen for his 1589 edition of *Magia naturalis* and its preface describing the natural philosopher "examining with lynx-like eyes the things that manifest themselves so that having observed them, he may zealously put them into operation." Francesco Stelluti, one of the group's members, wrote that the academy's purpose was to "penetrate into the inside of things in order to know their causes and the operations of nature that work internally, just as it is said of the lynx that it sees not just what is in front of it, but what is hidden inside" (Fig. 2.4).[62]

Returning to Rome, Cesi wrote glowingly of Naples and its dynamic scientific culture. In a letter to Stelluti, the young prince described Naples as a "paradise of delights, full of amenities and pleasures, relaxed in its loveliness, most beautiful, gentle, and charming; the abode of fertile Ceres, abundant Neptune, and most pleasing and courteous Venus".[63] During his visit Cesi was introduced to the wider Della Porta circle, including Giambattista's older brother Giovan Vincenzo, a versatile naturalist who was said to be "divine in astrology, both in the theoretical part that

[58] Della Porta 1957, 204–5; Gassendi 1657, 36–7.

[59] Findlen 1994, 228.

[60] Daston 1988.

[61] Piccari 2007, 34–43.

[62] Gabrieli 1935, 507.

[63] Olmi 1987, 33; Freedberg 2002, 245.

Fig. 2.4 Title page of Giambattista Della Porta, *Magia naturalis*, with a lynx pictured at the top

measures the motions of the stars and the practical part that foretells future events".[64] Cesi also met other members of the Della Porta circle, naturalists and mathematicians that gathered in the Della Porta family palace in Naples. It was a cosmopolitan group that greeted scholars from across Europe. Through Della Porta, Cesi got to know the talented botanist Fabio Colonna (1567–1640) and the astronomer and publisher Colantonio Stigliola (1546–1623), an adherent of the theories of Copernicus and Giordano Bruno.[65] After his initial visit, Cesi was determined to expand the Lincei to Naples and make the Della Porta circle its focus.

On its face Cesi's attraction to Della Porta, the consummate courtly magus-scientist, seems improbable. Cesi hated court life. It's impossible to be a true philosopher in the courts, he said, because the courtier's sole interest is to curry the prince's favor. "It is extremely dangerous for the philosopher to fall from his honored station into such a treacherous place full of parasites, buffoons, and flatterers".[66] Yet Cesi knew Della Porta's worth in cultural and intellectual credit. The world famous magus had a name to conjure with, a name that would bring luster to his newly formed society. He also had a large library and an impressive cabinet of curiosities, which Cesi hoped he would bequeath to the academy.[67] Flattering the old magus, Cesi offered Della Porta the position of head of the Lincei's newly formed Neapolitan branch.

During his visit to Naples (the first of several), Cesi caught a glimpse of a collection of manuscripts that would set Neapolitan scientific culture on fire. The work had caused a stir when it arrived in Naples and soon word got around Italy of its presence in the city. In 1589, Della Porta received a letter from the Bologna naturalist Ulisse Aldrovandi inquiring about the album, which he'd heard contained pictures and descriptions of the plants and animals of the New World. Della Porta replied that he'd seen the manuscript, attributing it to a certain "Cortes," obviously confusing the real author, Francisco Hernández, with the conquistador Hernán Cortés.[68] Hernández, King Philip II's protomédico, or chief health officer, had spent seven years in New Spain collecting, describing, and illustrating plants and animals. The king had sent Hernández to the New World with instructions to assemble all the information he could find out about Mexican natural history. When he returned to Spain in 1577 Hernández brought back a huge, disordered body of pictures and descriptions, and began the laborious process of organizing them and preparing the work for publication. When Hernández died in 1587, his task unfinished, Philip entrusted the manuscripts to his Neapolitan court physician Nardo Antonio Recchi, instructing him to "put them into order." When Recchi returned to Naples in 1589 with his redaction of Hernández's work, word got around immediately among the Neapolitan naturalists. The work that stirred such intense excitement and anticipation in Naples—indeed all of Europe—was a monumental collection of drawings

---

[64] Badaloni 1960, 679.

[65] Gabrieli 1929; Badaloni 1985.

[66] Cesi 1969, 68.

[67] Olmi 1987, 38.

[68] Gabrieli 1989, 731.

and descriptions of New World flora and fauna, the likes of which Europeans had never seen.[69]

In his letter to Aldrovandi Della Porta reported that Recchi had selected about 600 illustrations from the more than 4000 that Hernández had assembled, many done by indigenous artists. Ever the dramatist, Della Porta imagined that Hernández had died of sadness when the king's council determined that they were of little use because "they were of Indian plants that could not be used in Spain." More likely, the Spaniard died of exhaustion from the labor of his immense task. Though the Neapolitan naturalist community was astir about the Hernández texts, it was not easy to get a chance to view them. Recchi showed them to Della Porta, who judged them to be "rare, useful, and most extravagant."[70] But neither he nor Fabio Colonna, who also got to look at the texts, were able to pry them from Recchi's hands or convince the doctor to publish them. Imperato didn't fare much better. Yet once Cesi caught a glimpse of the illustrations he was determined to find a way to publish them under the auspices of the Lincean academy. The Linceans never stopped talking about the Hernández texts, as the process of publication dragged on. Although pieces of the massive work trickled out in various publications over the years, it was not until 1651 that Recchi's selection was finally printed in Rome in a striking edition under the title *Rerum medicarum Novae Hispaniae thesaurus*.[71]

Spanish Naples was the conduit through which news of the Hernández material, which would revolutionize natural history and world views, reached Europe. It was all completely new. Neither Pliny nor Aristotle nor any of the ancients had every described such things, plants and animals so new that they had no names but the Nahuatl terms that Hernández borrowed from his indigenous informants. Nothing like it had ever been seen. Time and again ancient authority collapsed before a single piece of evidence.[72]

Della Porta, Colonna, and Colantonio Stigliola all became members of the new branch of the Lincei. Imperato, however, wasn't invited to join. It's possible that Della Porta discouraged Cesi from inviting his rival, but more likely it was Imperato's questionable political involvements that caused Cesi to keep him at bay.[73] Imperato was deeply involved in the politics of the *popolo* party as captain of the people in the Piazza di Nido district of Naples. In Imperato's political and cultural circles, infused with Telesio's natural philosophy, reformers, naturalists, and politicians pressed questions about social justice and the possibility of reform.[74]

Such speculations quietly simmered in the academies until 1585, when they suddenly sparked open revolt. Resentment toward Spanish rule had deepened in the years following the outbreak of 1547. Deteriorating economic conditions and renewed hopes for independence from Spain rekindled revolutionary fervor. Though

[69] Chabran and Varey 2000.

[70] Gabrieli 1989, 732.

[71] Chabran and Varey 2000, 15–18.

[72] Pagden 1993, 91.

[73] Olmi 1987, 43; Stendardo 2001, 36.

[74] Marino 2002, 6.

seemingly incited spontaneously by an increase in bread prices, the insurrection of 1585 was no explosion of wild popular fury. It was a planned, popular revolution intended to overturn the social order. The revolt's leader, the pharmacist Giovan Leonardo Pisano, was well known in the political and social circles of the naturalists. His brother was Della Porta's philosophy tutor and he served, with Ferrante Imperato, as a captain of the *piazza*. People said that Pisano's laboratory (*officina*) was a meeting place for citizens "who made speeches against the viceroy".[75]

Ferrante Imperato's activism nurtured in his son Francesco a devotion to the people's cause. In 1598 Francesco Imperato, like his father a naturalist committed to the reform movement, published with Stigliola's press a treatise on the political organization of Naples, *Discorso politico intorno al reggimento delle piazze della città di Napoli*.[76] The detailed, empirical work, which amounted to a natural history of the Neapolitan political world, was one of the key texts radicals used in constructing the revolutionary ideology leading up to the revolt of Masaniello in 1647.[77] Both his book and his father's were natural histories, one of the natural world and the other of the political order. As Rosario Villari observes, the *Discorso* "sprang from an environment which was responsible for all the revolutionary efforts and ideas of reform of the closing years of the century."[78] While Ferrante Imperato and his son Francesco pushed the naturalists' reform agenda to its limits, Cesi rigorously avoided the taint of politics. He was adamant that his society be neutral in politics and religion. He knew its survival depended on it.

Although aspirations for sweeping religious and political reform were present in southern Italy since the early sixteenth-century, in reality there was little possibility of real change. As the apparatus of the Church and state tightened its grip over Italy, hopes for reform found expression in the urge to create alternate worlds, imaginary and better than the real world.[79] Inspired by legends of a lost golden age, rumors about noble savages in the New World, and popular tales of Coccagna, the mythical land of plenty, the imaginary utopias evoked images of a society close to nature and governed by simple rules of equity and justice. Southern Italy, imbued with the ideas of philosophical naturalism, was fertile soil for utopian fantasies. In Naples and Calabria, utopianism almost always took expression in the idea of a return to nature, or of creating new worlds by following nature's example. Most of the utopias were imaginary. Only one intellectual, Tommaso Campanella, a brilliant and erratic Dominican friar from Calabria, dared try to build one. Campanella's failed 1599 revolt, in which a paltry band of rebel clergy and nobles plotted to free Calabria and Naples from Spanish rule, was doomed to failure from the start. The revolt's ideology makes it unmistakably clear that Campanella had been indoctrinated by debates in the Neapolitan academies. In fact, Campanella admitted during his trial, it was during his first visit to Naples and from the Della Porta group that he came to

[75] Villari 1993, 29.

[76] Imperato 1604.

[77] Villari 1967, 108–9.

[78] Villari 1993, 66. See also Stendardo 2001, 38–49.

[79] Martines 1980, 322–31; Firpo 1957.

understand that there would be an imminent "mutation in human affairs" (*mutationes rerum humanarum*).[80]

Cesi's strategy of courting Della Porta and avoiding politics paid off. The Accademia dei Lincei survived its first crisis, buying time until 1611, when a new star, Galileo Galilei, appeared on the horizon. In publishing his *Siderius Nuncius* (Starry Messenger), Galileo became an instant scientific celebrity. Once Cesi persuaded Galileo to join the Lincei, the society's fortunes changed. Galileo's electrifying discoveries with the telescope gave the Lincei a new sense of purpose. Almost immediately, Galileo's influence on the society began to eclipse that of Della Porta. Abandoning its dilettantish preoccupation with secrets, the Accademia dei Lincei committed itself to the Galilean cause. Esotericism gave way to identification with the "republic of letters" and secrets no longer seemed to hold relevance. Della Porta's style of science—reading the book of nature as if it were written in hieroglyphics—didn't die out. In fact, it grew in stature and prominence in the Spanish Baroque, where it fit comfortably. Spanish intellectuals regarded Della Porta as the most Spanish of the Neapolitans. But the Lincei chose a different star to follow. Under the shadow of Galileo, who described the book of nature as being written in the language of mathematics, the idea that nature closeted deep and meaningful secrets seemed passé. As one scientific star slowly receded, a new one rose. Before long, the Lincei were committed to Galileo's research program.[81]

Stories of scientific stardom are fixtures of Renaissance Italian scientific culture. Scientific celebrity was invented in the Renaissance. It was largely a product of the printing press, but court culture also played a role. Leonardo Fioravanti managed to gain entry into Don Pedro of Toledo's court through contacts with the Toledo family that he made in Sicily. He served as a military surgeon in the imperial navy during the African campaign, where he perfected the balsam for treating wounds that would propel him into the limelight. He left Naples in 1555, a few years after meeting Della Porta, and went to Venice, where he became a celebrated author renowned for his "new way of healing." Milking the legend of Alessio Piemontese, he elevated himself in the popular press as the model professor of secrets: an innovative, crafty, untiring experimenter driven to discover the secrets of nature.[82]

In Venice, Fioravanti met a mathematician whose magic, he thought, exceeded even that of the wizard of Naples. He was Ettore Ausonio, a mathematics teacher who hailed from Milan.[83] Ausonio was an adept in the art of catoptrics, the branch of optics dealing with the formation of images by mirrors. Catoptrics has a rich history that stretches back to Archimedes and during the Renaissance was one of the subjects that made up the expansive field of natural magic. Describing Ausonio's inventions Fioravanti wrote of onlookers "stupefied" by their ingenuity and capacity to produce marvelous effects:

---

[80] Eamon 1995, 396; Villari 1993, 61–2.

[81] Eamon 1994, 231.

[82] Eamon 2010.

[83] Dupré 2005; Reeves 2008.

> He has made concave mirrors of impressive size in which marvelous and great things appear, which by now all the princes of the world know about them. One is so amazing that anyone who sees it is stupefied. I've seen him make so many things that produce strange and awesome visions that even if I were to try to describe all the bizarre things I saw it would be endless and by comparison to what he has done, nothing at all.[84]

Optical illusions of the sort that Della Porta performed before enthralled spectators were among the most celebrated achievements of natural magic. Although mirrors and lenses were familiar artifacts of urban life, mirrors that produced wondrous visions were novel and rare. Anamorphosis—the art of projecting distorted or "monstrous" images that become visible only when viewed through a curved mirror—was decidedly avant-garde in the sixteenth century.[85]

In 1576, Fioravanti followed in Della Porta's footsteps, making a pilgrimage to the court of Philip II, but his experience there was completely different. He became immersed in the court's smoke-filled alchemical world, with its swarms of adepts, Italians as well as Spaniards, busying themselves before alchemical furnaces that Philip had installed at the royal garden at Aranjuez to produce medicines and cosmetic waters for the court.[86] Fioravanti's quest for the *Magna Medicina* drove him to use alchemy to make daring, powerful, and dangerous drugs. After botching a cure and killing a patient, who happened to be a powerful courtier's servant, he was expelled from Castile. He returned to Naples in 1577, chastened but unapologetic. He disappears from the historical record after that, but in all probability he died in Naples. Years of alchemical experimenting in poorly ventilated rooms took their toll. Prolonged exposure to heavy metal vapors may, in fact, account for the instability, paranoia, and depression that he exhibited in his last writings.

Leonardo Fioravanti was the most famous professor of secrets of the age. He earned his fame by his skillful manipulation of the printing press. He was an accomplished writer and a deft reader of his audience. His picareque tales of miraculous cures and daring surgeries, travels to exotic places, and repeated battles with the medical establishment enthralled readers. And he had an uncanny ability to explain complex alchemical techniques in language that was concrete and tied to the familiar. His drugs, mainly purgatives calculated to produce dramatic results, brought him luster and made him the de facto head of an alternative medical movement. Fioravanti was a new kind of scientific celebrity. Compared to Della Porta, who occupied a lofty perch in the intellectual sphere, Fioravanti stood on a lower rung of the social ladder of professors of secrets, midway between the magus and the charlatan. Della Porta, Imperato, Fioravanti, and the street corner charlatan all made their art and performed their experiments on a stage, whether in Della Porta's elegant Villa delle Pradelle, in Imperato's museum, in the tales of scientific adventure that Fioravanti told in his books, or on the charlatan's makeshift stage in the piazza. All of these venues were spaces for the performance of science in early modern Naples, and were integral components of the city's scientific culture.

---

[84] Fioravanti 1567, 55.

[85] Eamon 2010.

[86] Rey Bueno 2009.

Della Porta craved fame but knew fame's downside. In 1613, two years before his death, he complained in a letter to Cesi that admirers came to visit him in droves, and that for every visitor bent on philosophical enlightenment there were four or five attracted by his reputation as a wizard. Although he claimed to be irritated by the crowds, he would willingly oblige visitors with a demonstration of one sort or another.[87] He'd become a carnival act, entertaining guests with experiments and tricks, and grew weary of it.

The renowned Spanish writer Francisco de Quevedo (1580–1645) was one of Della Porta's guests during his later years. Quevedo was only thirty-one at the time but his poetry and satire had drawn praise from writers such as Lope de Vega and Cervantes. Quevedo visited Della Porta around 1611 while serving as a counselor to the Sicilian Viceroy, the Duke of Osuna. The crafty old magus and the brash young poet probably met at one of the meetings of the recently founded Oziosi Academy (The Idlers). In contrast to the academies of secrets that were closed down by Toledo, the Oziosi were sponsored by the Spanish Viceroy, Pedro Fernandez de Castro, the Count of Lemos. The presence of the vice regal court attracted leading Spanish writers and intellectuals. Besides Quevedo, the playwright Félix Lope de Vega y Carpio (1562–1635) and the brothers Lupercio (1559–1613) and Bartolomé (1561–1631) Leonardo de Argensola, jointly known as "El Horacio de España," were all at one time or another members of the Oziosi.[88]

By the time of Quevedo's visit Della Porta was one of the most famous intellectuals in Europe and the greatest magus of the day. In addition to the phenomenally successful *Natural Magic*, he had published on optics, cryptology, agronomy, fortifications, human physiognomy, among other subjects. In one of his late works, *De aeris transmutationibus* (1610) the magus drew a parallel between distillation and meteorology. Della Porta regarded air as vehicle of celestial influences to earth, making meteorology a foundation of natural magic.[89] But the correspondence could have been with any of the sciences that Della Porta wrote about: all were interconnected. The wizard of Naples had opinions about witchcraft too, but he prudently left them unpublished.[90]

Quevedo writes that while residing in Naples he visited Della Porta in his private museum and was shown an *aposento de espejos*, or mirror chamber, that projected multiple images of the viewer, who "sees only what the mirrors represent." This would not have been one of the toy models of an "amphitheatrical glass" that Della Porta described in *Natural Magic*, but a room, an *aposento*—in this case a circular chamber with mirror lined walls (Fig. 2.5).

It's remarkable that Della Porta should have constructed such a space in his town house in Naples, where Quevedo visited him. Given the cost of mirrors it must have been an extravagant expenditure, one calculated to amaze the viewer. One can only imagine the conversation that took place during the visit. Seeing Della Porta's mir-

[87] Clubb 1965, 52.

[88] Green 1933.

[89] Borrelli 2011.

[90] Della Porta 1982; Piccari 2007, 54–8.

Fig. 2.5 Model of a mirror chamber, from Giambattista Della Porta, *Magiae naturalis libri XX* (1651)

ror chamber so impressed Quevedo that he remembered it years later when composing his pious and deeply philosophical *Providencia de Dios* (1641). To this baroque intellectual, Della Porta's mirror chamber was a metaphor standing for what happens "to the soul in the enclosed body, where imagination is surrounded by mirrors that display images, simulacra, and forms without matter"[91]

The infinite proliferation of apparitions in Della Porta's mirror chamber reminded Quevedo, a zealous champion of the Counter-Reformation, of the vanity of thinking about things that are impossible to understand—such as matters of faith. "How many times," he asks, "does the intellect ponder that which is nothing, will be nothing, and is impossible?" Like the observer in Della Porta's mirror chamber, the mind goes nowhere when it concerns itself with things it cannot know. Quevedo's visit to Della Porta's *museo* reminds us that the entire political and cultural world of late Renaissance Naples lay under the double shadow of Imperial Spain and the Counter-Reformation. These twin influences account in large measure for the peculiar character of Renaissance Neapolitan scientific culture.[92]

Quevedo's reflection on Della Porta's "*aposento de espejos*" brings us back to the metaphor of a "theater of experiments" that I've tried to elaborate in this essay. A theater of experiments isn't just a space where experiments are enacted and tricks performed. It's also a stage where nature performs. It's true that many of Della

[91] Quevedo 1958, 1411. "Esta dependencia accidental y concomitante te la asimilo al hombre que en un aposento de espejos no ve sino lo que los espejos le representan; y no obstante que el ojo que ve, no puede verse a sí, ni el uno al otro, ni los dos el aspecto donde están, sin el reflejo. – no por eso la potencia visiva es el reflejo, ni depende dél por sí, sino condicional y accidentalmente. Lo mismo sucede al alma, en el cuerpo cerrado, donde la imaginación la cerca de espejos, que la muestran imágines, simulacros y formas sin materia." [This accidental and concomitant dependence I compare to a man in a chamber of mirrors (such as the one I saw in the house of Giambattista Della Porta in Naples, a man learned in curiosities) who sees only what the mirrors represent to him; nevertheless the seeing eye cannot see itself or each other, nor [can] the two see the appearance where they are, without the reflection. - Not for that reason the visual faculty is a reflection, or depend on itself, but rather conditional and accidentally. The same happens to the soul, in the closed body, where the imagination is surrounded by mirrors that show images, simulacra, and forms without matter.]

[92] I am grateful to Professor John Slater for his insightful comments on the Quevedo passage.

Porta's experiments were essentially magician's tricks meant to deceive the viewer. But they were also serious scientific work, and were premised on deeply held beliefs about nature that are utterly foreign to the modern sensibility, including the assumption that nature performs, jokes, and sometimes even plays tricks. Nature's jokes (*lusus naturae*) are nature's way of expressing itself in the conversation of science. Nature's jokes aren't obvious to everyone, though. It takes a lot of erudition and keen observation to get them. Instances of the occult properties of things, nature's jokes are understood only with the discerning eye of the naturalist or the skillful hand of the magus. As Paula Findlen explains, *lusus* "became an organizing principle that described the process of diversification in nature".[93] Nature's diversity is a result of nature's self-fashioning, on display in the courtly play at work when the magus performs.

In a recent article, Arianna Borrelli argues that the distinctive characteristic of Della Porta's natural philosophy was its humanism—and by this she does not mean Renaissance humanism in the usual sense, a program for the revival of classical literature and philosophy, but rather the belief that "natural phenomena were primarily defined by how they looked, sounded, felt and acted upon those who witnessed them. For Della Porta, a natural philosopher should pursue his quest of natural secrets by asking: If I were nature, how would I go about producing this effect?"[94] Producing it, one might add, in an appropriate conversational setting, such as the Villa delle Pradelle, for an audience of curious gentlemen versed in the classics.

This depiction of Della Porta's experimental methodology eloquently captures the magus's style of doing science, and aligns neatly with the metaphor of a theater of experiments. Quevedo's reflection on Della Porta's *aposento de espejos* takes the metaphor further, suggesting that a theater of experiments is several things at once. First it's a space where a play is enacted and tricks performed, a space for revealing and skillfully concealing secrets; second, a space for demonstrating experiments before an audience of curious virtuosi; third, a stage where nature plays, joking and reenacting the stories of Ovid and Pliny; and finally, a metaphorical space, where the search for the wider meaning of nature's secrets begins. All were aspects of natural magic's humanism and all were enacted in the theater of experiments that defined Renaissance Neapolitan scientific culture.

More than just histrionic, however, Della Porta's experiments were demonstrations and reminders of the principles of natural magic, which taught that nature is alive and sentient and packed with powerful occult forces to which only the learned magus had access, that nature performs and in performing declares itself and provides parables for living, and that sensation and emotion are traits of nature. Natural magic made perfect sense to Neapolitans. Living in the shadow of Vesuvius and upon the Campania's fecund volcanic soil, they were present to the primeval forces of nature that natural magic tapped into. They knew nature's power. Telesio provided Neapolitan intellectuals with a metaphysics that made sense of the nature they

[93] Findlen 1990.

[94] Borrelli 2011.

knew, and Della Porta gave them tools for mining its secrets. Beyond the scientific, naturalism gave intellectuals a blueprint for self-fashioning and a rationale by which to liberate Naples from Spanish dominance. Neapolitan political culture nourished naturalism and natural magic, and they in turn shaped Neapolitan scientific culture.

References

Agrimi, Mario. 1992. Telesio nel seicento Napolitano. In *Bernardino Telesio e la cultura napoletana*, ed. Raffaeli Sirri and Maurizio Torrini, 332–372. Naples: Guida editori.

Albertus Magnus, Pseudo 1973. *The book of secrets of Albertus Magnus of the virtues of herbs, stones and certain beasts*, ed. Michael R. Best and Frank H. Brightman. Oxford: Oxford University Press.

Aquilecchia, Giovanni. 1968. Appunti su G. B. Della Porta e l'Inquisizione. *Studi secenteschi* 9: 3–31.

Badaloni, Nicola. 1960. I fratelli Della Porta e la cultura magica e astrologica a Napoli nel '500. *Studi Storici* 1: 677–715.

———. 1980. Fermenti di vita intellettuale a Napoli dal 1500 alla metà del '600. In *Storia di Napoli*, vol. 8: 307–47. Naples: Edizioni Scientifiche Italiane.

———. 1985. Il programma scientifico di un bruniano: Colantonio Stigliola. *Studi Storici* 26: 161–174.

———. 2004. *Inquietudini e fermenti di libertà nel rinascimento italiano*. Pisa: ETS.

Borrelli, Arianna. 2011. Giovan Battista Della Porta's Neapolitan magic and his humanistic meteorology. In *Variantology 5. On deep time relations of arts, sciences and technologies*, ed. Siegfried Zielinski and Eckhard Fürlüs, 103–130. Cologne: Walther König.

———. 2014. Thinking with optical objects: Glass spheres, lenses and refraction in Giovan Battista Della Porta's optical writings. *Journal of Early Modern Studies* 1: 39–61.

Brown, Jonathan, and John H. Elliott. 1980. *A palace for a king: The Buen Retiro and the court of Philip IV*. New Haven: Yale University Press.

Canepa, Nancy L. 2013. Literary culture in Naples, 1500–1800. In *Companion to early modern Naples*, ed. Tommaso Astarita, 427–451. Leiden: Brill.

Cañizares-Esguerra, Jorge. 2006. Chivalric epistemology and patriotic narratives: Iberian colonial science. In *Nature, empire, and nation: Explorations of the history of science in the Iberian World*, 7–13. Stanford: Stanford University Press.

Castiglione, Baldesar. 1959. *The book of the courtier*. Trans. Charles S. Singleton. Garden City: Anchor Books.

Cesi, Federigo. 1969. Il natural desiderio del sapere. In *Scienziati del seicento*, ed. Maria Luisa Altieri Biagi, 53–92. Milan: Rizzoli.

Chabrán, Rafael, and Simon Varey. 2000. The Hernandez Texts. In *The Mexican treasury: The writings of searching for the secrets of nature. The life and works of Dr. Francisco Hernández*, ed. Simon Varey, 3–25. Stanford: Stanford University Press.

Clubb, Louise George. 1965. *Giambattista Della Porta, Dramatist*. Princeton: Princeton University Press.

Cocco, Sean. 2013a. Locating the natural sciences in early modern Naples. In *Companion to early modern Naples*, ed. Tommaso Astarita, 453–76. Leiden: Brill.

———. 2013b. *Watching Vesuvius. A history of science and culture in early modern Italy*. Chicago: University of Chicago Press.

Cummins, Stephen. 2015. Encountering Spain in early modern Naples: Language, customs and sociability. In *The Spanish presence in sixteenth-century Italy: Images of Iberia*, ed. Piers Baker-Bates and Miles Pattenden, 43–61. Farnham: Ashgate Publishing.

Daston, Lorraine. 1988. The factual sensibility. *Isis* 79: 452–467.
Daston, Lorraine, and Katharine Park. 1998. *Wonders and the order of nature, 1170–1750*. New York: Zone Books.
de Quevedo, Francisco. 1958. *Obras completas I: Obras en prosa*, ed. Felicidad Buendia. Madrid: Aguilar.
De Young, Ursula. 2011. *A vision of modern science: John Tyndall and the role of the scientist in Victorian culture*. London: Palgrave MacMillan.
Della Porta, Giambattista. 1558. *Magiae naturalis, sive de miraculis rerum naturalium libri IIII*. Naples: M. Cancer.
———. 1589. *Magiae naturalis libri viginti*. Naples: H. Salvian.
———. 1957. *Natural Magick*. New York: Basic Books (orig. publ. 1658).
———. 1980. *Gli duoi fratelli rivali* (ed. and trans. Louise George Clubb). Berkeley: University of California Press.
———. 1982. *Criptologia*, ed. Gabriella Belloni. Edizione Nazionale di Classici del Pensiere Italiano, ser. 2, 37. Rome: Centro Internazionale di Studi Umanistici.
Dupré, Sven. 2005. Ausonio's mirrors and Galileo's lenses: The telescope and sixteenth-century practical optical knowledge. *Galilaeana. Journal of Galilean Studies* 2: 145–180.
Eamon, William. 1994. *Science and the secrets of nature: Books of secrets in medieval and early modern culture*. Princeton: Princeton University Press.
———. 1995. Natural magic and utopia in the Cinquecento: Campanella, the Della Porta Circle, and the revolt of Calabria, *Memorie Domenicane*, n.s., 26: 369–402.
———. 2010. *The professor of secrets: Mystery, medicine and alchemy in Renaissance Italy*. Washington: National Geographic.
Eamon, William, and Françoise Paheau. 1984. The Accademia Segreta of Girolamo Ruscelli. A sixteenth-century Italian scientific society. *Isis* 75: 327–342.
Everson, Jane E. 2012. The melting pot of science and belief: Studying Vesuvius in seventeenth-century Naples. *Renaissance Studies* 26: 691–727.
Findlen, Paula. 1990. Jokes of nature and jokes of knowledge: The playfulness of scientific discourse in early modern Europe. *Renaissance Quarterly* 43: 292–331.
———. 1994. *Possessing nature. Museums, collecting, and scientific culture in early modern Italy*. Berkeley: University of California Press.
Fioravanti, Leonardo. 1567. *Dello Specchio di scientia universale*. Venice: A. Ravenoldo.
Firpo, Luigi. 1957. *Lo stato ideale della controriforma*. Bari: Laterza.
Freedberg, David. 2002. *The eye of the Lynx. Galileo, his friends, and the beginnings of modern natural history*. Chicago: University of Chicago Press.
Gabrieli, Giuseppe. 1929. Intorno a Nicola Antonio stigliola, filosofo e Linceo napoletano. *Giornale critico di filosofia italiana* 10: 469–485.
———. 1935. Spigolatura Dellaportiane. *Rendiconti della R. Accademia Nazionale dei Lincei. Classe di scienze morali, storiche e filologiche*, ser. 6, 11: 491–517.
———. 1989. *Contributi all storia della Accademia dei Lincei*. Rome: Accademia Nazionale dei Lincei.
Garzoni, Tomaso. 1996. *La piazza universale di tutte le professioni del mondo*, ed. Paolo Cherchi, 2 vols. Turin: Einaudi.
Gassendi, Pierre. 1657. *The Mirrour of true nobility & gentility. Being a life of the renowned Nicolaus Claude Fabricius Lord of Peiresk, Senator of the Parliament at Aix* (trans. W. Rand). London: J. Streater for Humphrey Moseley.
Gliozzi, Mario. 1950. Sulla natura dell' 'Accademia de' Secreti' di Giovan Battista Porta. *Archives Internationales d'Histoire des Sciences* 12: 536–541.
Green, Otis. 1933. The literary court of the Conde de Lemos at Naples, 1610–1616. *Hispanic Review* 1: 290–308.
Hernando Sánchez, Carlos José. 1994. *Castilla y Nápoles en el Siglo XVI. El Virrey Pedro de Toledo: Linaje, estado y cultura: (1532–1553)*. Junta de Castilla y León, Consejería de Cultura y Turismo.

Imperato, Francesco. 1604. *Discorso politico intorno al reggimento delle piazze della città di Napoli*. Napoli: Stigliola.

Kamen, Henry. 1997. *The Spanish Inquisition. An historical revision*. London: Phoenix Press.

Kodera, Sergius. 2012. Giambattista Della Porta's histrionic science. *California Italian Studies* 3 (1), http://escholarship.org/uc/item/5538w0qd.

———. 2014. The laboratory as stage: Giovan Battista Della Porta's experiments. *Journal of Early Modern Studies* 3: 15–38.

Lopez, Pasquale. 1974. *Inquisizione stampa e censura nel regno di Napoli tra '500 e '600*. Naples: Edizioni del Delfino.

Marino, John. 1982. The 'Trickster Economy' in the Kingdom of Naples. *Comparative Studies in Society and History* 24: 211–234.

Marino, John A. 2002. On the shores of Bohemia. In *Early modern history and the social sciences: Testing the limits of Braudel's Mediterranean*, John A. Marino, 3–32. Truman State University Press.

———. 2011. *Becoming Neapolitan. Citizen culture in Baroque Naples*. Baltimore: Johns Hopkins University Press.

Martin, John. 1996. The imaginary piazza: Tommaso Garzoni and the late Italian Renaissance." In *Portraits of medieval and Renaissance living: Essays in memory of David Herlihy*, ed. Samuel Kline Cohn, 439–454. Ann Arbor: University of Michigan Press.

Martines, Lauro. 1980. *Power and imagination. City-states in Renaissance Italy*. New York: Vintage Books.

Minieri-Riccio, Camillo. 1878–1880. Cenno storico delle accademie fiorite nella città di Napoli. *Archivio storico per le province napolitane*, 3 (1878): 746–58; 4 (1879): 163–77, 379–94, 516–36; 5 (1880): 131–57, 349–73, 578–612.

Moran, Bruce T. 2005. *Distilling knowledge: Alchemy, chemistry, and the scientific revolution*. Cambridge, MA: Harvard University Press.

Muraro, Luisa. 1978. *Giambattista Della Porta mago e scienzato*. Milan: Feltrinelli.

Musi, Aurelio. 2007. The kingdom of Naples in the Spanish imperial system. In *Spain in Italy: Politics, society, and religion 1500–1700*, ed. Thomas James Dandelet and John A. Marino, 73–98. Leiden: Brill.

Olmi, Giuseppe. 1981. 'In essercitio universale di contemplatione, e prattica': Federico Cesi e i Lincei. In *Università, accademie e società scientifiche in Italia e in Germania dal Cinquecento al Settecento*, ed. L. Boehm and E. Raimondi, 169–236. Bologna: Il Mulino.

———. 1987. *La colonia lincea di Napoli. Atti del Convegno Galileo e Napoli*, 23–59. Naples: Guida.

Pagden, Anthony. 1993. *European encounters with the New World*. New Haven: Yale University Press.

Paniccia, Ersilia. *Francesco Imperato. Napoli e la rivoluzione di Masaniello*. Archigrafica. www.archigrafica.org/imperato.pdf.

Pereira, Michela. 1990. Lullian Alchemy: Aspects and problems of the corpus of alchemical works attributed to Ramon Lull (XIV–XVII century). *Catalan Review*: 41–54.

Piccari, Paolo. 2007. *Giovan Battista Della Porta: il filosofo, il retore, lo scienziato*. Milan: FrancoAngeli.

Reeves, Eileen. 2008. *Galileo's glassworks. The telescope and the mirror*. Cambridge, MA: Harvard University Press.

Rey Bueno, Mar. 2007. Los destiladores reales de los Austrias españoles (1564–1700). *Azogue. Revista electrónica dedicada al estudio histórico-crítico de la alquimia* 5: 108–129.

———. 2009. *La mayson pour distiller des eaües* at El Escorial: Alchemy and medicine at the court of Philip II, 1556–1598. *Medical History* 53(Suppl S29): 26–39.

Rey Bueno, Mar and María Esther Alegre Pérez. 2001. Los destiladores de Su Majestad. Destilación, espagiria y paracelsismo en la corte de Felipe II. *Dynamis: Acta Hispanica ad Medicinae Scientiarumque Historiam Illustrandam* 21: 323–350.

Riskin, Jessica. 2008. Amusing physics. In *Science and spectacle in the European Enlightenment*, ed. Bernadette Bensaude-Vincent and Christine Blondel, 43–60. Aldershot: Ashgate.
Romeo, Giovanni. 2013. Inquisition and Church in early modern Naples. In *Companion to early modern Naples*, ed. Tommaso Astarita, 235–256. Leiden: Brill.
Silverblatt, Irene. 2004. *Modern Inquisitions: Peru and the colonial origins of the civilized world.* Durham: Duke University Press.
Stendardo, Enrica. 2001. Ferrante Imperato: collezionismo e studio della natura a Napoli tra Cinque e Seicento, Napoli, Accademia Pontaniana. Quaderni dell'accademia pontaniana, vol. 31.
Tasso, Torquato. 1982. *Tasso's dialogues* (Trans. Carnes Loard and Dain A. Trafton). Berkeley: University of California Press.
Van Deusen, Neil C. 1935. The place of Telesio in the history of philosophy. *The Philosophical Review* 44: 417–434.
Varchi, Benedetto. 1827. *Questiones sull' alchimia*, ed. D. Moreni. Florence: Magheri.
Villari, Rosario. 1967. *La rivolta antispagnola a Napoli. Le origini (1585–1647).* Bari: Laterza.
———. 1993. *The revolt of Naples*. Cambridge: Polity.
Wistreich, Richard. 2007. *Warrior, courtier, singer: Giulio Cesare Brancaccio and performance of identity in the late Renaissance*. Aldershot: Ashgate.

Chapter 3
Giambattista Della Porta: A Magician or an Optician?

Yaakov Zik and Giora Hon

Abstract In Bk. 17, Ch. 4 of *Magia Naturalis* (1589) Giambattista Della Porta (ca. 1535–1615) reported his experiments on concave spherical mirrors arranged in various setups. Della Porta identified two critical points: (1) the point of inversion (*punctum inversionis*) in reference to the place where the magnified image is turned upside down and seen blurred, and (2) the point of burning (*punctum incensionis*) in reference to the place where the reflected rays concentrate and ignite fire. Opticians and practitioners of the time distinguished between the two points but considered them to occupy the same spatial location.

Della Porta inferred from his studies of concave spherical mirrors that the position of the point of inversion and that of the point of burning occupy different spatial locations. He associated the point of inversion with a locus where the image is seen magnified, turned upside down and blurred—a matter of visual perception. He defined the point of burning as a physical, optical position associated with a geometrical point in which the converging rays ignite fire. Consequently, throughout Bk. 17, Della Porta discarded the point of inversion from his optical nomenclature and referred only to the point of burning, the real—so to speak—optical point. In so doing, Della Porta contributed fundamentally towards the technological management of sets of optical elements.

In this paper we follow the experimental practice of Della Porta as presented by the optical demonstrations in Bk. 17, Ch. 4. We discuss the theoretical principles Della Porta developed to clarify whether his claim concerning concave spherical mirror is hypothetical or was it based on an inference from experience. We offer novel insights into the development of the theory of reflection in concave spherical mirrors as it was pursued by Della Porta. He eliminated perceptual considerations from his optics and considered only geometrical-physical aspects. This approach was most useful in the development of the telescope where the critical aspect is not perception but rather ratio of spatial angles.

Y. Zik (✉) • G. Hon
Department of Philosophy, University of Haifa, 31905 Haifa, Israel
e-mail: zikya@013.net; hon@research.haifa.ac.il

A. Borrelli et al. (eds.), *The Optics of Giambattista Della Porta (ca. 1535–1615): A Reassessment*, Archimedes 44, DOI 10.1007/978-3-319-50215-1_3

3.1 Introduction

In Bk. 17, Ch. 4 of *Magia Naturalis Libri XX* (1589) Giambattista Della Porta (ca. 1535–1615) reported his experiments on the functioning of concave spherical mirrors arranged in various setups.[1] He portrayed wide variety of wonders and visual effects that could be generated by mirrors and identified two points: (1) the point of inversion (*punctum inversionis*) in reference to the position where the image seen in concave spherical mirrors (and convex lenses) achieves its greatest magnification just before it turns upside down as the eye is moved further away from the optical element; and (2) the point of burning (*punctum incensionis*) in reference to the physical point at which light rays were concentrating by reflective and refractive elements.

In canonical optical texts of the time, e.g., Ptolemy (ca. AD 90–ca. AD 168), Alhacen (ca. 965–ca. 1040), Roger Bacon (ca. 1220–ca. 1292), John Pecham (ca. 1230–ca. 1292), and Witelo (ca. 1230–ca. 1280), the geometrical nomenclature used for referring to the point of reflection (*punctus reflexionis*) in mirrors, or the point of refraction (*punctus refractionis*) in refractive elements, was related to the points on the surface where an incident ray is broken. Della Porta associated the point of reflection (*punctum reflexionis*) with the point where the reflected ray meets the optical axis of the mirror, that is, the point of burning. Della Porta associated the point of incidence (*punctum incidentiae*) with the point where the incident ray strikes the surface of the mirror.[2]

Of particular interest for us is the work of Girolamo Cardano (1501–1576). In his *De Subtilitate Libri XXI* ([1550] 1580) he outlined the properties of a concave spherical mirror with a reference to his authoritative source, namely, Euclid.[3] According to Cardano, Euclid correctly demonstrated (*rectè demonstrauit*) that when a concave spherical mirror is placed out against the sun, the reflected rays are directed to the center of the mirror onto a single point (*vnum punctum*) where fire is kindled.[4] Cardano explained that at the point where the rays of the sun come together:[5]

1. The force (*vis*) of the reflected rays generate heat. The bigger the diameter of the mirror is, the more and faster it ignites fire. Yet such a mirror cannot ignite fire at a distance since the rays are always concentrated in its center.
2. The reflected rays illuminate (*illuminant*) the neighboring parts of that point, and
3. The images (*imaginem*) are seen confused and turned upside down.

---

[1] Della Porta (1589, Bk. 17: 264–265).

[2] Della Porta (1593, Bk. 2: 39–41).

[3] In addition to Cardano, Della Porta also referred to, among others, Alhacen, Archimedes, Aristotle, Euclid, Galen, Pecham, Ptolemy, Vesalius, and Witelo as sources for his knowledge of optics.

[4] Cardano ([1550] 1580, Bk. 4: 164, 165, 166); Forrester (2013, Bk. 4: 244, 246). Cardano acknowledged the merits of a parabolic mirror for directing the rays emanating from the Sun to one point: see Cardano ([1550] 1580, Bk. 4: 166); Forrester (2013, Bk. 4: 246–248).

[5] Cardano ([1550] 1580, Bk. 4: 149, 165–166, 170); Forrester (2013, Bk. 4: 222–223, 245–246, 251).

Like Cardano, opticians and practitioners of the time distinguished between the point of inversion and the point of burning but considered them to occupy the same spatial location.[6]

Della Porta inferred from his studies of concave spherical mirrors that the place of the point of inversion and that of the point of burning occupy different spatial locations. This discovery is critical; its importance cannot be exaggerated. Indeed, this is the principal claim of our paper. Della Porta was well familiar with the circumstances in which the image seen in concave spherical mirrors and convex lenses depends upon the relative locations of the eye, the object, and the optical element. The image can be of the same size as, larger than, or smaller than, the object depending on its relative position. The image can also appear reversed or upright as a function of the positioning of the eye in front or behind the point of inversion. Della Porta associated the point of inversion with a locus—not a geometrical point—where the image seen by the eye is turned upside down as a function of the placement of the object and the eye in relation to the optical element. He defined the point of burning as a physical, optical place associated with a geometrical point at which the converging rays ignite fire. Consequently, throughout Bk. 17, Della Porta discarded the point of inversion from his optical analyses and referred *only* to the point of burning, the real—so to speak—optical point which is independent of the perceiving eye. Della Porta determined the location of this point as a function of the mirror's radius curvature; the point of burning is located on the optical axis at a distance equals to $^{1}/_{4}$ of the mirror's diameter.[7] For this optical analysis Della Porta referred the reader to his other optical book, that is, *De Refractione Optices Parte Libri Nouem*, which he technically published in 1593 but evidently the study was available to the public earlier than this date of publication.[8]

In *De Refractione* Della Porta explained, *inter alia*, the novel geometrical method he developed for determining the burning point of concave spherical mirrors as a function of the mirror's radius curvature and the height of the incident ray.[9] While Della Porta's method facilitated practical applications of the law of reflection,[10] which was well established towards the end of the sixteenth century, the principles governing refraction were a mystery.[11] In *Magia Naturalis* and *De Refractione* Della

[6] Della Porta ([1558] 1562, Bk. 4, Ch. 14: 124–125); Smith (2015: 344). See also Dupré (2005: 152).

[7] Della Porta (1589, Bk. 17: 265, 271).

[8] Della Porta (1589, Bk. 17: 271); Della Porta (1593, Bk. 2: 40–41). On Della Porta's experiments with optical objects and his optical writings, see Borrelli (2014, 1: 39–61).

[9] Della Porta (1593, Bk. 2: 36–41).

[10] On the law of reflection, see Smith (2015: 55–59).

[11] In Bk. 2 of *De Refractione*, Della Porta introduced the reciprocity between reflection and refraction as the framework for his study of refraction. For that purpose Della Porta explored and physically illustrated phenomena of reflection and refraction in mirrors and glass spheres which he thought could illustrate the visual faculties. See Zik and Hon (2012: 445–456); Smith (2015: 347–349).

Porta noted that plane concave and convex mirrors, as well as non-equal transparent bodies (e.g., lenses, prisms), deceive in delivering the images.[12]

We discuss Della Porta's widely circulated optical works assuming the following historiographical presuppositions: (1) Della Porta should be read and judged in the context in which he was working, and (2) Bk. 17 of *Magia Naturalis* and *De Refractione* should be read as one body of optical knowledge. Della Porta made ample cross references in the two books.

We first review the historical data related to the questions:

1. Did Della Porta experiment on reflection in mirrors, or was his practice merely illustrations of phenomena? And,
2. Was Della Porta's claim concerning concave spherical mirror hypothetical or was it based on an inference from his experimental studies?

Against the historical data (Sect. 3.2), we make two fundamental claims: (1) Della Porta developed a novel practice for experimenting on concave spherical mirrors, and (2) this practice facilitated quantitative demonstration of the law of reflection for concave spherical mirror. We then follow the "optical footprints" in *Magia Naturalis* and *De Refractione*.

In order to substantiate these fundamental claims we follow Della Porta's experimental practice as it is exhibited in the optical demonstrations in Bk. 17, Ch. 4 (Sect. 3.3). We present optical simulations to illustrate the phenomena Della Porta wished to demonstrate in his experiments on concave spherical mirrors (Sect. 3.4). We discuss the theoretical principles Della Porta developed to clarify whether his claim concerning concave spherical mirror is hypothetical or was it based on an inference from experience (Sect. 3.5)? In conclusion (Sect. 3.6), we discuss the merits of Della Porta's theory of reflection in concave spherical mirror.

3.2 Review of the Facts

We begin by recording the relevant historical data as they appear in Bk. 17 and Ch. 4 of *Magia Naturalis*:

1. From the outset of Bk. 17 and up to the end of Ch. 3, Della Porta demonstrated the operations of a single as well as arrays of plane mirrors.
2. In Ch. 4 Della Porta turned to the diverse optical effects of concave spherical mirrors. Right from the beginning of the chapter he emphasized the need to know the position of the point of inversion, which he then associated with the point where the reflected rays of the sun are united.
3. Della Porta then demonstrated the operations of concave spherical mirrors in relation to the point of inversion, and portrayed the diverse effects they produced.

[12] Della Porta (1589, Bk. 17: 259–261); Della Porta (1593: 1–6, Bk. 8: 173).

4. At the end of Ch. 4 Della Porta introduced the point of burning into his analysis, that is, the point where the converging rays create the most powerful brightness and heating capacities and ignite fire. From this juncture to the end of Bk. 17 Della Porta did not discuss any more the point of inversion; he referred only to the point of burning.
5. At the end of Ch. 6 Della Porta made the first of 7 references in which the reader is encouraged to consult his other optical book, that is, *De Refractione.*[13]
6. In Ch. 22 of Bk. 17 Della Porta provided technical instructions of how plane and spherical mirrors of good quality are produced.[14] Della Porta's account was not merely theoretical; he owned a workshop for producing optical elements.[15] This is evident from Francesco Fontana's (1580–ca. 1656) attempts to get possession of the instruments for lens making left after the death of Della Porta in 1615.[16]

These historical data serve as a background to our discussion.

3.3 Della Porta's Experimental Practice

In Bk. 17, Chap. 4 of *Magia Naturalis*, titled, *On the diverse effects of concave mirror*, Della Porta demonstrated visual effects that can be generated by concave mirror. At the outset of the chapter he emphasized the importance of knowing the location of the point of inversion, that is, the place where the image is seen turned upside down. Della Porta demonstrated how:[17]

1. **To know the place of the point of inversion of images in a concave mirror:** One should set the mirror against the sun and where one sees the rays united there the point of inversion is located.[18]
2. **To see things magnified:** One should set one's head below the point of inversion, closer to the surface of the mirror, where one could see one's face huge and monstrous and one's finger as great as one's arm. At that point "women could pull hairs of their eyebrow for they are seen as large as finger."[19]

---

[13] Della Porta (1589, Bk. 17: 267): "In nostris in opticis fusius declaratum est."

[14] Della Porta (1589, Bk. 17, Ch. 22: 279). Note that Della Porta's technical instructions given in chapters 17–21, and 23, include detailed account for the production of parabolic mirrors, optical elements of cylindrical shape, spectacles lenses, and looking mirrors made of glass or metal: see Della Porta (1589, Bk. 17: 275–280).

[15] On Della Porta and his optical experience in Venice, see Reeves (2008: 70–78).

[16] Crasso (1666, Vol. 2: 297).

[17] Della Porta (1589, Bk. 17: 264–265).

[18] In effect this is the place where the point of burning (focal point) of the mirror is located.

[19] Della Porta (1589, Bk. 17: 264).

3. **To kindle fire with a concave glass:** One should set a mirror, made of excellent glass, against the sun so that the converging rays will form a brilliant pyramid of light. Then one could place any combustible matter in the center of the mirror where it will kindle fire, melt lead or tin, and make gold or iron red hot.[20]
4. **To see the image hanging in the air:** One should set one's head further away from the point of inversion where the image will appear inverted far from the mirror. At the place where the cathetus[21] cuts the line of reflection, the image is seen almost separated from the mirror. The nearer one is to the center (point of inversion) the larger is the image seen as if one could touch it with one's hand. The larger the diameter of the mirror is, one might see more parts of the object. And if a man is running toward the mirror with a drawn sword one may see another man runs to meet one's self. If one looks at a candle, one would think it is pendulous light in the air.
5. **To see the image far from the center:** When one moves one's eye further away from the point of inversion the image is seen inverted and less magnified and the picture of a thing (object) shall be farther stretched forth. If one moves the eye at that place to the right or left-hand toward the edges of the mirror, the picture of things would be seen stretched forth about the mirror surface at the point where the cathetus touches the line of reflection. In so doing "many strange wonders, which few persons were able to reproduce, may be observed."[22]
6. **To reflect heat, cold, and voice:** If a candle is placed where the visible object is to be set, the image of the candle will come to one's eyes and offend them with heat and light. When one puts snow at that place the eye will presently feel the cold. But there are greater wonders in it; for it will not only reflect heat and cold, but voice too, and make echo. This is demonstrated by placing a man in front of a mirror while another person stands far behind his back. When the man approaches closer than the point of inversion at the right hand of the mirror, his face will be seen magnified far from the mirror. Now, when the man is speaking with low voice against the mirror, the other person could hear the words and see the motion of the mouth of the reflected image.
7. **To see in the night what is done afar:** One should place a candle at the point of inversion so that the incident rays, from the center of the mirror to its circumference, which are reflected in parallel paths, will cast the light to the desired place even if it is located far from the mirror. Any object placed between the reflected parallel rays will be clearly illuminated.
8. **To illuminate a great hall with a few and small lights:** One should set concave mirrors at two or more places at a distance (from the illuminated area) where their reflected parallel rays may be coincident. Then lamps should be

---

[20] Here Della Porta was still considering the point of inversion and the point of burning occupying the same place at the center of the mirror.

[21] On the cathetus rule of image formation, see Smith (2015: 59–62).

[22] Della Porta (1589, Bk. 17: 264).

placed in the respective points of inversion of the mirrors so that the reflected light is multiplied as if it is noonday. One, who knows the craft, could mark letters with wax or some dark substance on the surface of the mirror and set it against the multiplied illumination so it casts the light upon the walls of a darken chamber where he could easily read the letters.

9. **To read letters in the dark night:** One should set a mirror against Venus or Mercury, or against a fire or light that is far off.[23] The reflected rays will meet at the burning point (*incensionis puncto coibit*) and cast a most bright light whereby one could easily read the smallest letters. For if the point of reflection (*punctum reflectit*), that is, the point of burning, is aimed at illuminating every word one could see the characters clearly.
10. **To set a house or fort on fire:** One should place the mirror against the sun so that the converging rays may fall upon the point of burning. Then one could lay fuel and things that will take fire or heaps of gunpowder if one wished to blow up towers. One could hide the mirror and at the next day the sun will fall upon the same point where the combustion materials for the fire have been placed.

We conclude that the optical demonstrations with concave spherical mirrors, as reported by Della Porta, were not merely illustration of phenomena; in effect, they attest to a systematic experimental practice.

We claim that Della Porta could not perform these demonstrations successfully without being aware of the following facts:

1. The image can be of the same size as, larger than, or smaller than its object, depending on the distance of the eye from the mirror (e.g., demonstrations 2, 4, and 6, above).
2. The image appears upright or reversed as a function of the position of the eye in front or behind the point of inversion (e.g., demonstration 4, above).
3. The location where the image is seen depends upon the relative displacements of the eye, the object, and the optical element (e.g., demonstrations 4, 5, and 6, above).
4. The diameter of the mirror has no effect on the magnification. The larger the mirror's diameter is, the more parts of the object can be seen (e.g., demonstration 4, above).
5. Concave mirrors have the capacities to project heat, cold, and voice (e.g., demonstration 6, above).
6. There are differences in locations and magnitudes of projected images and images seen by the eye (e.g., demonstrations 6, 7, and 8, above).
7. The point of inversion and the point of burning are not located at the same place (e.g., demonstrations 9, and 10, above).

[23] Della Porta (1589, Bk. 17: 265).

Della Porta discovered that the point of inversion is associated with visual perception while the point of burning is a physical, optical location associated with a geometrical point at which the converging rays ignite fire. Della Porta then discarded the point of inversion from his optical discussion in Bk. 17 and referred only to the point of burning. His formulation is quantitative. Already in the first edition of *Magiae Natvralis* Della Porta had acknowledged that important correlation exists between the segment of the circle, that is, the region bounded by a chord and the arc subtended by the chord, of the concave mirror and the distance at which it kindles fire.[24] He claimed throughout Bk. 17 that all the demonstrations, either with plane or spherical mirrors and lenses, would fail if knowledge of the exact radius of the circles from which the optical elements are configured was not available. This knowledge is crucial for properly setting the correct distances between the optical elements used in his optical demonstrations.[25] Della Porta further stressed that practitioners did not have this knowledge and yet undertook to teach it, and uttered therefore nothing but nonsense. He emphasized that none before him was able to obtain this knowledge.[26] The replacement of a subjective notion, the point of inversion, with an objective locus, the point of burning, constitutes a fundamental contribution towards the technological management of sets of optical elements.

3.4 Optical Simulations of Della Porta's Demonstrations

To study the feasibility of the optical experiments Della Porta presented in the course of his investigations in Chapter 4 of Bk. 17, we apply the following simulations, made with a concave spherical mirror.[27] Each simulation reproduces effects resulted from specific setup of the mirror, the location of the object, and the placement of the eye at various distances from the mirror in reference to the point of inversion and the focal point of the mirror. It also reproduces the appearance of the image in terms of its magnitude, blurriness (denoted by the hatched area), and whether it is turned reverse or upright.

[24] Della Porta ([1558] 1562, Bk. 4, Ch. 14: 125): "Si verò maioris sphæræ fuerit segmentum, per maiorem accendit distantiam."

[25] Della Porta (1589, Bk. 17: 266): "Id tamen duximus ad monedum, neoperam frustreris, quod proportionati sint oportet circuli specilli & concaui portio."

[26] Della Porta (1589, Bk. 17: 267): "Nec leues poterunt imaginari technae, distantiam speculi magnitudine emendabis. Sat habes, qui id docere conati sunt, non nisi nugas protulere, necaliquibus ad huc compertum putarim."

[27] The optical properties of the mirror we used in the simulations are: radius curvature 54 cm, diameter 15 cm, and focal length 27 cm. For measuring the distances among the elements in the test we used an optical bench. The computations throughout this paper were made with *OSLO* (Optical Software for Layout and Optimization).

Simulation 1: The appearance of the image when the eye and the object (e.g., a candle) are being gradually moved further away from the mirror. Near the mirror the image is seen upright and minified. Further away it is magnified and getting blurred nearer to the point of inversion. At the center of curvature, where the point of inversion is located, the image is most magnified, reversed, and blurred. Then, with the increasing range, the image is seen reversed, less blurred, and minified.

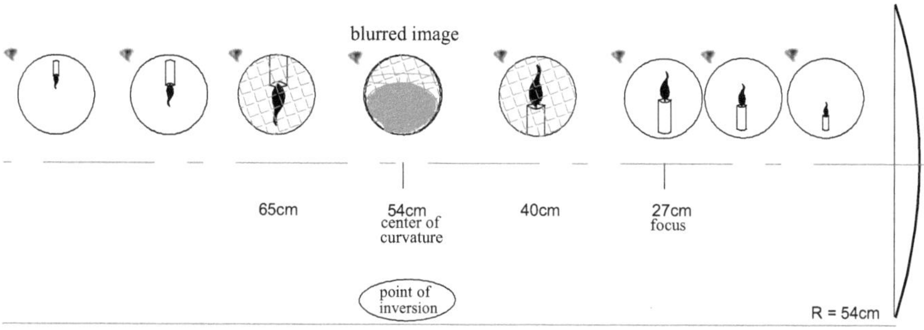

Simulation 2: The appearance of the image when the object is placed at the center of curvature and the eye is being gradually moved further away from the mirror. Near the mirror the image is seen upright and minified. Further away it is magnified and getting blurred nearer to the point of inversion. At the center of curvature, where the point of inversion is located, the image is most magnified, reversed, and blurred. Then, with the increasing range, the image is seen reversed, less blurred, and minified.

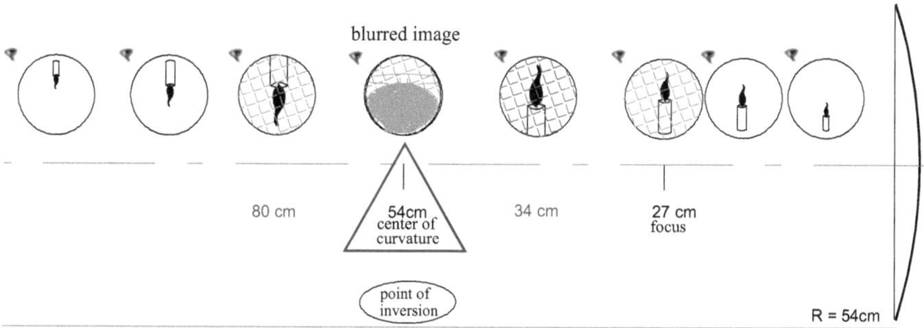

Simulation 3: When the object and a screen are placed at the center of curvature the image is depicted on the screen inverted at the same magnitude. If the eye is replacing the screen the image is seen most magnified, inverted, and blurred.

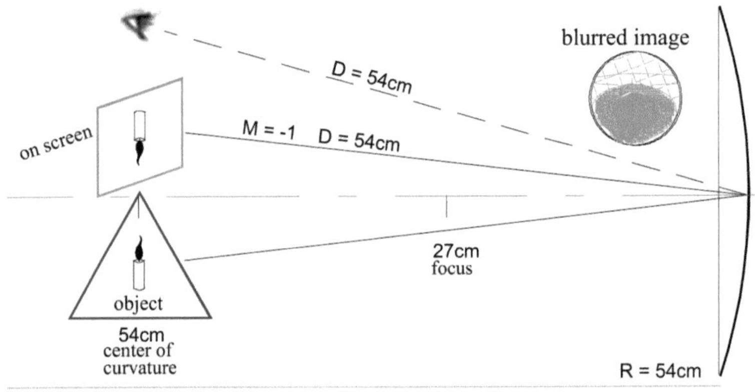

Simulation 4: When the object is placed about 600 cm from the mirror and the eye is being gradually moved further away from the mirror the point of inversion is located at 28 cm from the mirror. Near the mirror the image is seen upright, a bit magnified and blurred. At the point of inversion the image is most magnified and blurred. Further away the image is seen clearer, minified, and reversed.

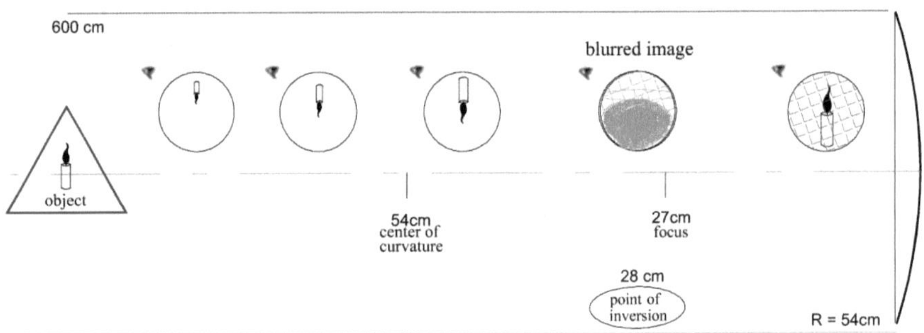

Simulation 5: When the eye is placed about 110 cm from the mirror and the object is being gradually moved further away from the mirror the point of inversion is located 37 cm from the mirror. The image is upright and minified near the mirror and getting larger and blurred towards the point of inversion. Further away the image is reversed, getting clearer, and minified.

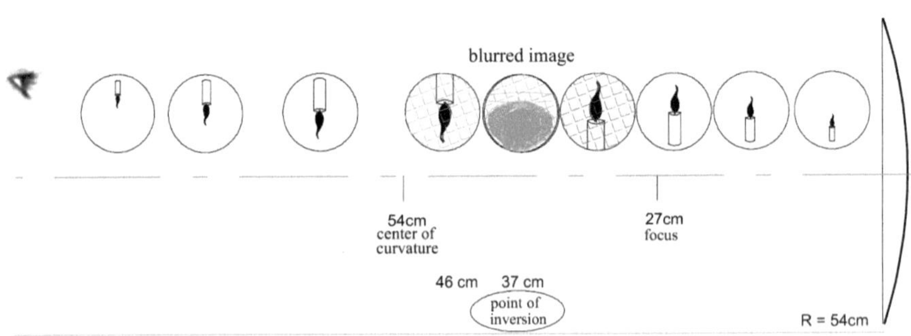

Simulation 6: When the object is placed about 37 cm from the mirror and the eye is being gradually moved further away from the mirror the point of inversion is located at 85 cm. Near the mirror the image is seen minified and upright and getting larger and more blurred towards the point of inversion. Further away the image is seen reversed and then it becomes clearer and minified.

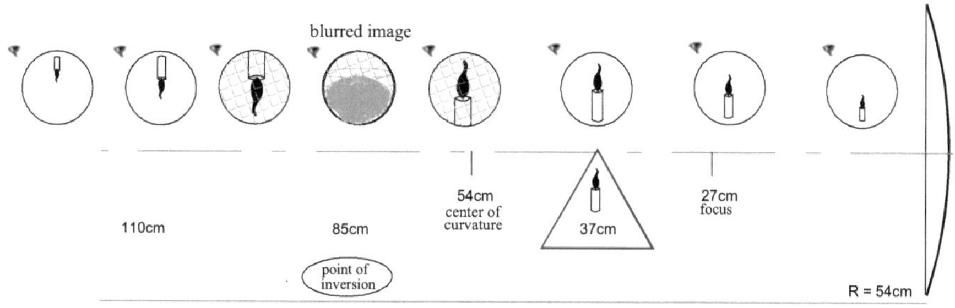

Simulation 7: The projection capacities of concave spherical mirror depends on the relative locations of the object and the screen. When a candle is placed at a distance of 29 cm from the mirror, the image would be clearly depicted inverted and magnified (–) 13.5 times at a distance of 392 cm. When the candle is placed at a distance of 35 cm from the mirror, the image would be clearly depicted on screen inverted and magnified (–) 3.75 times at a distance of 118 cm. Finally, as shown in the upper part of the simulation, when the candle is replaced by a bright source of light placed near to the focal point of the mirror at a distance of 27.2 cm from the mirror, the image would be clearly depicted on screen inverted and magnified (–) 135 times at a distance of 36.7 m.

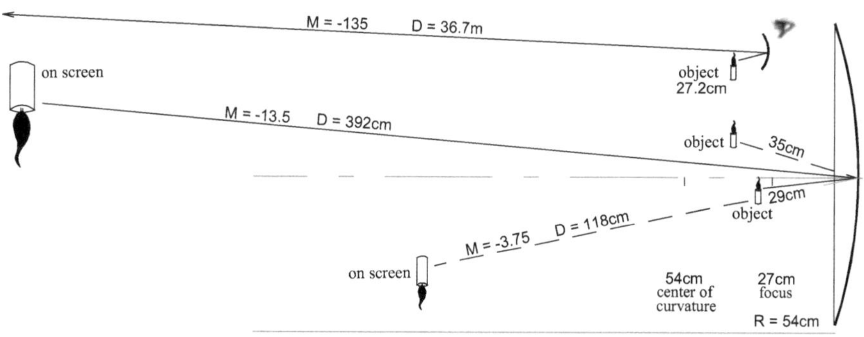

Simulation 8: We now turn to the illumination capacities of the mirror. As shown in the center of the simulation, when a source of light such as a candle, is placed at the point of inversion located 54 cm from the mirror, a clear patch of light is cast at a distance of about 60 cm from the mirror. In the upper part of the simulation we see that when the candle is placed at the point of inversion located 28 cm

from the mirror, the patch of light is cast at a distance of 760 cm from the mirror. A candle placed at the point of inversion located 37 cm from the mirror casts the light at a distance of 120 cm from the mirror while the candle placed at the point of inversion located 85 cm from the mirror, does not cast any illumination.

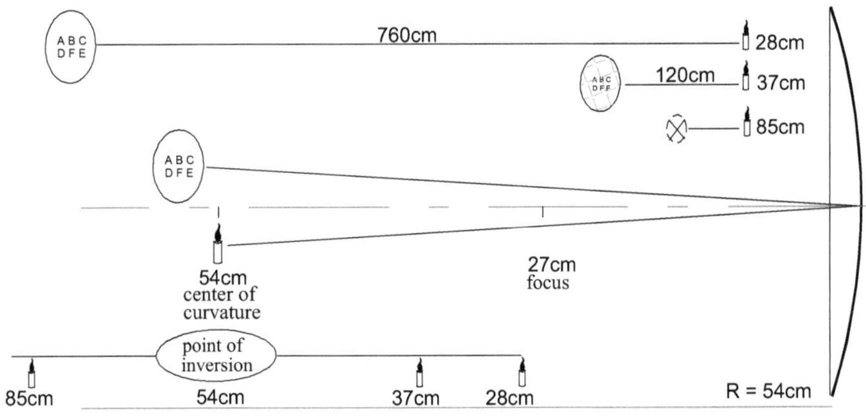

The simulations clearly show that:

1. The place of the point of inversion is not fixed. It varies according to the relative positions of the object and the eye.
2. The point of inversion and the point of burning do not occupy the same spatial location.
3. The larger the mirror's diameter is, the more parts of the object are seen, that is, the field of view is increased.
4. The diameter of the mirror does not affect its power.

The optical simulations support the feasibility of Della Porta's experimental practice and the validity of his discoveries.

3.5 Della Porta's Optical Theory

Della Porta's methodological approach and quantitative reasoning are well described in his optical writings. At the outset of *Magiae Naturalis* and *De Refractione* Della Porta characterized optics as mathematical knowledge (*mathematicas scientias*) in which catoptrical experiments (*experimenta* e) combined with mathematical demonstrations could provide definitive conclusions. He thought that geometrical speculations supported by observation should be considered true and formulated the following rules which were accepted by authors of optics (*perspectiuae authoribus accepta*): (1) Some of the rays emanating from the sun are parallel to one another; (2) The rays correspond to straight lines in geometrical demonstrations; and (3)

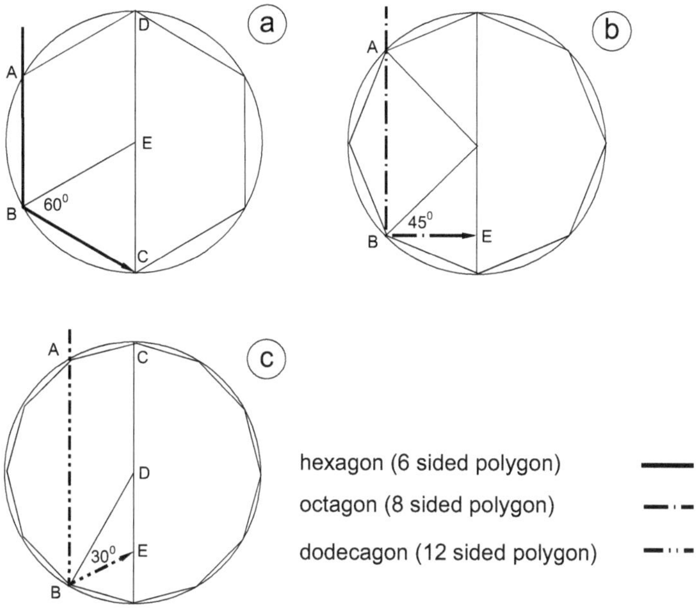

Fig. 3.1 Reflection in hexagon, octagon, and dodecagon

Solar rays incident on the surface of convex, concave, or plane mirrors always form equal angles of incidence and reflection.[28]

In accordance with this methodological position, Della Porta initiated in *De Refractione* a geometrical study of the relation between the incident and reflected rays in concave spherical mirrors. He referred the reader to the optical demonstrations with concave spherical mirrors he made in *Magia Naturalis*.[29] Della Porta concluded this study by formulating a general proposition for the relation between the point of incidence, namely, the point where the ray is reflected, and the point of reflection, the point where the reflected ray intersects the optical axis of the mirror for any incident ray. Della Porta showed geometrically that the point of reflection of a concave spherical mirror is located on the optical axis at a distance equals to half the radius curvature of the mirror.[30] This quantitative result was put forward in *Magia Naturalis*.[31]

Della Porta built his analysis on Euclid's *Elements*, Bk. IV, in which the geometrical properties of polygons inscribed in a circle are discussed.[32] He began the discussion: Let DABC (Fig. 3.1a) be a concave spherical mirror to which a ray of

[28] Della Porta (1593, Bk. 2: 35–36). See also Della Porta (1593, Bk. 2: 3–6) and Della Porta (1589, Bk. 17: 259–260). On Renaissance art, naturalism, and optics, see Smith (2015: 298–321).

[29] Della Porta (1593, Bk. 2: 40): "Vt libro naturalis Magiae demonstrauimus."

[30] Della Porta (1593, Bk. 2: 39, 41).

[31] Della Porta (1589, Bk. 17: 271).

[32] See, Densmore (2002, Bk. 4: 83–98).

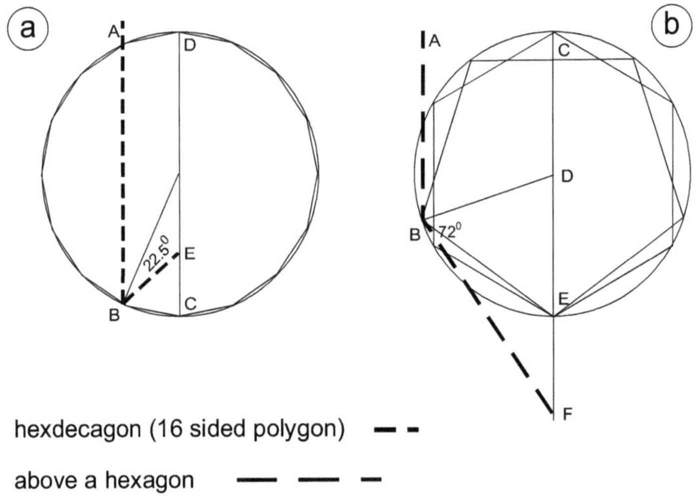

Fig. 3.2 Reflection in hexdecagon, and above hexagon

the Sun AB, parallel to the diameter CED, incident at B, and arc BC equals to one sixth of the circle circumference. The connected lines BC, CE, and EB form a triangle BCE in which angles CEB, EBC, and BCE are equal and the alternate angles CEB and ABE are also equal. Since the angle of incidence is equal to the angle of reflection, the incident angle ABE is equal to the angle of reflection CBE, and line BC is the path of the reflected ray. Therefore, in a spherical concave mirror, when a ray of the Sun, AB strikes along the chord of a hexagon, it is reflected towards the diameter DC at its lower terminus to point C (60°).

On the basis of these relations Della Porta inferred the following propositions: the angle of reflection, made by a hexagon (Fig. 3.1a, 6 sided polygon) is 60°; the angle of reflection made by an octagon (Fig. 3.1b, 8 sided polygon) is 45°; the angle of reflection made by a dodecagon (Fig. 3.1c, 12 sided polygon) is 30°.

The angle of reflection made by hexdecagon (Fig. 3.2a, 16 sided polygon) is 22.5°; and so the angles of reflection are getting more acute as the rays strike the spherical concave mirror nearer to the diameter. Hence, the point of reflection (*punctum reflexionis*) for each ray, respectively, that is, point E, is located along the diameter CD from the terminus at C (Fig. 3.2a), up to a point placed at one quarter of the mirror's diameter, where the point of reflection (that is, the burning point) of the mirror is located.

Della Porta noted that by placing the burning point, that is, the focal point of the mirror, at one quarter of its diameter he corrected Euclid's long standing error in *Catoptrica*, where the burning point of a concave spherical mirror was placed at the center of the mirror's diameter.[33] Della Porta now traced the reflection when ray AB (Fig. 3.2b) strikes above the chord of a hexagon at a height equal to a chord of a

[33] Della Porta (1593, Bk. 2: 39).

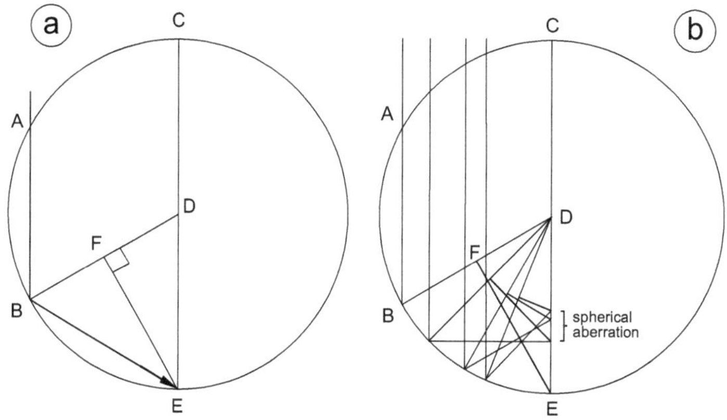

Fig. 3.3 The determination of the point of reflection and the caustic formed by reflection

pentagon; the angle of incidence ABD is equal to the angle of reflection DBF. The path of the reflected ray BF falls beyond the diameter CDE, further than E, to the point of reflection F outside the spherical concave mirror.

Della Porta concluded the discussion of reflection with a general proposition: To find the point of reflection for a given point of incidence (Fig. 3.3a): Let EBAC be a spherical concave mirror, B a point of incidence on its circumference. To find the point of reflection on the diameter CE, draw line DB from the center of the circle, D, to the point of incidence, B, and mark point, F, at the middle of that line. From point F, draw a perpendicular towards DE. Point E, where the perpendicular FE intersects the diameter, CE, marks the place where the point of reflection is located. Because in triangles FBE and FDE the angles DFE and BFE are right angles, and lines BF and FD are equal, angles FDE and FBE are also equal. Lines DE and BE are equal and the point of burning, which Della Porta also denoted as the point of reflection, is located at the place where the two lines coincided, namely, E.

By applying an equilateral polygons inscribed by a circle, Della Porta established a method by which the relations among, (1) the height of the incident ray denoted by its chord, (2) the angle of reflection, and (3) the radius of the mirror, could be calculated. Accordingly, the focal length of a concave spherical mirror could be determined in terms of the radius of the spherical mirror.

Della Porta's geometrical construction is applicable for any given incident ray and shows how the optical path of the reflected ray could be traced (Fig. 3.3b): The intersection of a perpendicular, dropped from the mid-line connecting the point of incidence and the center of a spherical concave mirror, and the diameter of the mirror, always marks the point of reflection. One could thus establish quantitatively the caustic formed by reflection.

In this optical study Della Porta made use of the point of incidence and the point of burning (also denoted as the point of reflection, namely, the focal point) which are geometrical points. The point of burning occupies a spatial location, but the point of inversion is a *locus* where the image is seen inverted—a matter of visual perception. This is not a measurable point! It shifts continuously and could be determined at several different locations. Della Porta associated the geometrical point of reflection with the burning point, that is, the focal point of the mirror. He then stopped referring to the point of inversion.

3.6 Conclusion: The Merits of Della Porta's Theory of Reflection

We followed the optical experiments with concave spherical mirrors which Della Porta presented in *Magia Naturalis.* We examined the logical consequence of the geometrical analysis he developed in *De Refractione*. Faithful to his methodological approach, Della Porta linked his experimental practice with geometrical demonstrations. The distinction between two optical points—the geometrical and the visual—convinced Della Porta of the importance to apply geometrical analysis to the optical experiments he had designed. His goal was to obtain new *quantitative* results and the geometrical analysis indeed delivered them.

The links Della Porta established between geometrical analysis and systematic procedures for reproducing optical phenomena attest to his scientific methodology—one needs to test in physical situations the relevant geometrical propositions. Della Porta's claim concerning concave spherical mirror was not hypothetical. He formulated a theoretical scheme based on inferences from experiments.

Della Porta's discoveries facilitated a new theory of concave spherical mirrors. His approach was quantitative. Using a straightedge, compass, and table of chords, he could determine geometrically optical properties of a concave spherical mirror in terms of the radius of curvature and the height of the incident ray. Given the radius curvature and the height of the incident ray, Della Porta could calculate the place of the focal point on the optical axis of the concave spherical mirror. This is an important conceptual move: Della Porta eliminated perceptual considerations from his optics and considered only geometrical-physical aspects. This approach was most useful in the development of the telescope where the critical aspect is not perception but rather ratio of spatial angles.

We are now in a position to answer the question, Was Della Porta a magician? Or was he an optician? We reply, he was both, for it is the context that determines the answer. Given the conception and practice of natural magic as an emerging idea of science in the latter half of the sixteenth century, he was a magician.[34] However, given his experimental work on the passage of rays in optical arrangements and the quantitative results which he obtained and shared in publications with his fellow practitioners, he was an optician, and a good one at that, we may add.

---

[34] On Della Porta's philosophy of natural magic, see Della Porta (1589: v–viii, Bk. 1: 3–4). See also Saito (2014: 17–39) and Daston and Park (2001).

Acknowledgments We gratefully acknowledge the helpful correspondence and ensuing discussions with A. Mark Smith and his critical comments related to the translation of Della Porta's *De Refractione*. This research is supported by the Israel Science Foundation (Grant No. 67/09).

References

Borrelli, A. 2014. Thinking with optical objects: glass spheres, lenses and refraction in Giovan Battista Della Porta's optical writings. *Journal of Early Modern Studies* 1: 39–61.

Cardano, G. [1550] 1580. *De Subtilitate Libri XXI*. Lvgdvni: Apud Stephanum Michaelem.

Crasso, L. 1666. Francesco Fontana. In *Elogii D'huomini Letterati Scritti*, 2 Vols, Vol. 2, 296–300. Venetia: Per Combi & La Noù.

Daston, L., and K. Park. 2001. *Wonders and the Order of Nature 1150–1750*. New York: Zone Books.

Della Porta, Giambattista. [1558] 1562. *Magiae Natvralis, Sive De Miracvlis Rervm Natvralivm Libri 4*. Antverpiae: In aedibus Ioannis Steelfij.

———. 1589. *Magia Naturalis Libri XX*. Neapoli: Apud Horatium Saluianum.

———. 1593. *De Refractione Optices Parte Libri Nouem*. Naples: Apud Io. Iacobum Carlinum and Antonium Pacem.

Densmore, D., eds. 2002. *Euclid's Elements*. New Mexico: Green Lion Press.

Dupré, S. 2005. Ausonio's mirrors and Galileo's lenses: The telescope and sixteenth century practical optical knowledge. *Galilaeana* 2: 145–180.

Forrester, J., eds. 2013. *The De Subtilitate of Girolamo Cardano*. Tempe: ACMRS.

Reeves, E. 2008. *Galileo's Glassworks*. Cambridge, MA: Harvard University Press.

Saito, F. 2014. Knowing by doing in sixteenth-century natural magic: Giambattista della Porta and the wonders of nature. *Circumscribere* 14: 17–39.

Smith, A. M. 2015. *From Sight to Light, the passage from ancient to modern Optics*. Chicago: The University of Chicago Press.

Zik, Y., and G. Hon. 2012. Magnification: How to turn a spyglass into an astronomical telescope. *Archive for History of Exact Sciences* 66: 439–464.

Chapter 4
Optical Diagrams as "Paper Tools": Della Porta's Analysis of Biconvex Lenses from *De refractione* to *De telescopio*

Arianna Borrelli

4.1 Introduction

During the European Renaissance mirrors and lenses made out of transparent, colourless and highly homogeneous "crystal" glass became available to an increasingly wide audience, prompting new developments in natural philosophical and natural magical research on light and vision.[1] There are clear indications that the new experiences and reflections also built upon artisan practices, although it is difficult for historians to grasp the exact extent and nature of this connection, since extant source are mostly of a textual character and were produced in a scholarly or courtly context.

Giambattista Della Porta was at the forefront of these developments and his writings are the earliest source providing evidence of systematic empirical study of the refractive properties of crystal glass spheres and lenses.[2] Whether Della Porta personally obtained all results presented or was reporting the outcome of collective experiences remains open, but in any case a special personal accomplishment of his was the development of methods for mediating through text and diagrams the knowledge produced by a combination of written and non-written practices. As I have argued elsewhere, Della Porta's aim in experimenting was not to search for

[1] Dupré 2000, 2003, 2005, Ilardi 2007, Lindberg 1984, Smith 2015, 322–372.

[2] On Della Porta's optical works see: Borrelli 2014, Lindberg 1984, Smith 2015, 344–349, Zik and Hon 2012 and references therein, as well as Zik and Hon's contribution to the present volume (Chap. 3). On his life, work and methodology see Borrelli 2011, Piccari 2007 and references therein.

A. Borrelli (✉)
History of Science, Technische Universität Berlin, Berlin, Germany

Institute for Advanced Study on "Media Cultures of Computer Simulation (MECS)", Leuphana University, Lüneburg, Germany
e-mail: aborrelli@weatherglass.de

A. Borrelli et al. (eds.), *The Optics of Giambattista Della Porta (ca. 1535–1615): A Reassessment*, Archimedes 44, DOI 10.1007/978-3-319-50215-1_4

universal laws of refraction, vision or light propagation. His main concern was to find practical rules for manipulating lenses and glass spheres to obtain specific, sensually perceivable effects, such as seeing an enlarged or inverted image, or lighting a fire.[3] If this plan worked out, the artefact became something more: an instrument whose rules of functioning provided means for manipulating and therefore understanding nature, conforming to the ideal of working knowledge characteristic of natural magic. In this search for the working principles of lenses and other refracting systems, diagrams were the main mediating instance between experience and reflection, as well as between the natural magician and his public, i.e. the audience of performances and the readers of books and manuscripts. In Della Porta's optical research diagrams took up an epistemic function like they did in the work of Kepler and Descartes, as convincingly argued by Albrecht Heeffer in his contribution to this volume,[4] but the way diagrams worked in the two cases was quite different, as we shall see.

In the last decades, the epistemic relevance of mediation and representation strategies in the construction of scientific knowledge has been demonstrated by a large number of studies in the history and philosophy of science.[5] It has become generally accepted that the bottom-up development of methods for representing and manipulating empirical results is a constitutive element of the process of their conceptualization: Words, symbols, formulas or diagrams on a page can act as "paper tools" to connect, reflect and expand instrumental and laboratory experience.[6] In the present contribution I will argue that optical ray-tracing diagrams took up such a function in Della Porta's optical writings. Moreover, I will suggest that Della Porta's methodology offered a very fruitful heuristic template for contemporary and later investigators of optical phenomena, despite the fact that the final results of these reflections not always corresponded to what later on would be regarded as the principles of geometrical optics. Thanks to diagrammatic practices, Della Porta was able to extend to lenses the connection between light focusing properties and visually perceivable effects that had been recently established for convex mirrors. Even in the case of mirrors the relationship was highly non-trivial, yet of paramount importance, as it allowed to extend the geometrical treatment of visual effects. At the core of the connection stood the ambiguous concept of "point of inversion", and we shall follow Della Porta's innovative attempts to adapt it to lenses by using diagrams as paper tools. While we today may regard the connection between a geometrically defined focal point and the phenomenon of image inversion quite straightforward, this is only the case because we conceive that effect already from the start in terms of an optical-geometrical diagram (i.e. the one for the "real" image), and not in terms of the visual inversion effect experienced when looking through a lens. However, the goal pursued by Della Porta, and by many of his contemporaries, was precisely an operative conceptualization of all such visually perceivable effects with

[3] Borrelli 2011, 2014.

[4] Heeffer 2017.

[5] See for example: Coopmans et al. 2014, Galison 1997, Gooding 1990, Kaiser 2005, Klein 2003.

[6] Klein 2003.

the aim of replicating and thus understanding them. This was a different epistemic goal from the formulation of general geometric laws of light and vision, and it was a goal which could not be attained by disregarding "subjective" vision. Because of this, the "point of inversion" remained a necessarily ambiguous notion, which Della Porta tried to represent and manipulate using diagrams as paper tools. In this context, a powerful diagrammatic tool was the cathetus line, which from the point of view of today's geometrical optics instead appears devoid of any physical meaning.

In the next section I will explain more in detail what a "paper tool" is and how diagrammatic methods contribute to knowledge production. In the following sections I will offer an overview of Della Porta's attempts to formulate general rules for constructing optical diagrams from which predictions of effects generated by spherical lenses could be derived. I will focus on the way in which he appropriated the methods of image construction in mirrors already available around the middle of the sixteenth century and applied them exploratively to biconvex lenses. Della Porta discussed optical phenomena in both editions of his "Natural magic" ("Magia naturalis", 1st ed. 1558, 2nd ed. 1589), in the treatise *De refractione* (1593, short: DR) and in the extant manuscript draft of an unfinished treatise known as *De telescopio* (ca. 1610–1615 short: DT).[7] Analysing how Della Porta's treatment of biconvex lenses developed in time, I will show how step by step Della Porta detached the cathetus line from its traditional function in perspectivist optics and employed it as a free-floating means of physical-geometrical reasoning to help devise diagrammatic rules for predicting under which circumstances lenses would produce effects like magnification or image inversion. Appropriating and transforming the cathetus line, Della Porta was able to build a bridge between the geometrically determined focal point of a lens on the one side and its visually defined point of inversion on the other, embedding the resulting hybrid concept in a set of diagrammatic methods as a parameter for determining whether image inversion and magnification would occur or not.

The plan of the paper is the following: In Sect. 4.2 I expound the notion of "paper tool" as introduced by historian Ursula Klein, and explain in what sense Della Porta's optical diagrams functioned as paper tools. Building upon previous scholarship dealing with Della Porta's analysis of spherical mirrors and the glass sphere, in Sect. 4.3 I will offer a short overview of knowledge about convex mirrors and lenses at the time Della Porta composed DR.[8] In Sect. 4.4 I will then focus on his treatment of biconvex lenses in chapter VIII of DR. A short general discussion of the manuscript draft of the treatise DT is given in Sect. 4.5: the draft contains Della Porta's later attempts to develop and expound a "geometrical optics" of lenses, as well as his explanation of the telescope. Della Porta never had time to put these reflections into a form fit for publication, and the manuscript has so far attracted limited interest

[7] I made use of the following editions of these works: Della Porta 1560, Della Porta 1597, Della Porta 1593, Della Porta 1962. The manuscript of DT is now in the Library of the Accademia dei Lincei in Rome and is dated to the last years of Della Porta's life (Della Porta 1610–1615)

[8] Borrelli 2014, Duprè 2006, Smith 2015, 280–391, Zik and Hon 2012.

on the part of historians. DT contains three versions of a chapter devoted to biconvex lenses (DT 3A, 3B, and 3C) and I shall discuss them in Sects. 4.6, 4.7, 4.8 and 4.9. I conclude in Sect. 4.10 by summarizing the thesis of the paper and the arguments I brought forward in its favour, and by setting my reflections in the context of scholarship on practices of mathematisation of experience in early modern and modern science.

4.2 Optical Diagrams and Their Epistemic Function as "Paper Tools"

Looking into mirrors, through lenses or in more complex optical apparatuses was in the early modern period an experience which could be widely shared, while at the same time retaining a subjective character. This was due not only to the different quality of apparently similar optical artefacts and to individual differences in sight strength, but also to the fact that what one would see through an optical device also depends on the way the observer is (or is not) trained to look through it. For example, the way in which today we look through a thin biconvex lens is informed by its predefined function as a "magnifying glass" and by the geometrical-optical rules for deriving what are seen today as its two main optical effects: the formation of a "virtual" magnified image and of a "real" inverted one. As we shall see more in detail later on, this distinction, while in many ways extremely useful, can become problematic when one tries to apply it to other effects produced by a lens, such as the small inverted image of an object located far away and looked at through a magnifying glass.[9] More in general, today a distinction is made between the laws governing reflection and refraction and the subjective impressions of what is seen in a mirror or through a lens. A particularly relevant example is the apparent location in three-dimensional space of an object seen through lenses. This effect is not usually discussed today among the basics of geometrical optics, but the apparent location is a constitutive element of the images seen in mirrors and lenses, and in the early modern period it was just as much an optical explanandum as inversion or magnification. Early modern practitioners of optics were confronted with the full spectrum of optical experiences with lenses and spheres without any codified means to represent and conceive them, and they tried their best to understand them in terms of the optical artefacts they were already familliar with: plane and spherical mirrors, and plane refractive surfaces. Such systematic attempts are recorded for the first time in Della Porta's writings and, as I have argued elsewhere, they can be characterized as "thinking with (optical) objects" in the same sense as the one defined by historian Domenico Bertoloni Meli (2006) for the case of early modern mechanics.[10] "Thinking with objects" means tentatively conceiving the behaviour of as yet unexplored

---

[9] I was not able to find any contemporary manual containing a geometrical-optical diagram illustrating how such images are formed.

[10] Bertoloni Meli 2006.

artefacts by conceptualizing them in terms of already known ones. In the case of optics, this was done by appropriating the techniques of ray-tracing developed to treat mirrors and plane refractive surfaces, and adapting them to deal with refraction in glass spheres and lenses. In the present contribution I will explore more in detail the epistemic significance of these diagrammatic practices, both as a part of a process of "thinking with objects", and in their own right.

All practitioners of optics in Della Porta's time made use of geometrical diagrams, but the Neapolitan scholar was particularly innovative in his approach, as he did not simply use them to study a small number of selected optical configurations, but rather employed them systematically to try and spell out rules allowing to schematically represent all possible effects that could be produced with a given artefact.[11] To this aim, he tried to list all such effects, independently of their more or less trivial character, and then strove to fit them all with the same procedures of diagrammatic representation. As we shall see, this modus operandi led him to become aware of similarities and differences which might have escaped a more theory-driven investigation. In discussing the various functions of diagrams in Renaissance optics, Sven Dupré argued that Della Porta's diagrams in DR had a "rhetoric" function, because they worked as "analogies" to establish a parallel between spherical mirrors and the glass sphere, and because they conveyed the impression that his explanations had the exactness of Euclidian demonstrations, while the diagrams did not "prove" anything in a strict logical-mathematical sense.[12]

While I thoroughly agree with Dupré on the fact that the diagrams played a key mediating role between mirrors and lenses, and that their function was in a sense "rhetorical", I also believe and will endeavour to show that diagrams constituted a fundamental part of Della Porta's methods of knowledge production. They were indispensable tools not only to support, but also to conceive optical arguments. At the same time, I shall not claim that Della Porta's diagrams were rigorous tools providing Euclidian proofs, as might be the case with authors like Francesco Maurolico or Descartes.[13] Della Porta's diagrams indeed had in so far a rhetorical function, as they served to let readers grasp the working of lenses according to the rules Della Porta had spelled out, and at the same time convince the audience of the validity of such rules. The diagrams fulfilled this function on the one side by defining the structural features shared by a potentially infinite number of individual cases, and on the other by demonstrating the empirical adequacy of the results achieved by assuming the validity of those common features. The latter justification was in turn based on a comparison between the diagrams and the phenomena which the readers had been instructed to "see" in the lenses, and therefore its validity

[11] Borrelli 2014.

[12] Dupré 2006, 33–38. Dupré's analysis is partly biased by an incorrect reading of Della Porta's text and drawings, which was in turn largely based on the equally incorrect interpretation of the same passages given by David Lindberg (1964). See Borrelli 2014 for a detailed discussion of this issue.

[13] The function of optical diagrams in those two authors are discussed in Heeffer's (Chap. 7) and Bellé's (Chap. 8) contributions to the present volume.

depended on the general, verbal and non-verbal context in which the diagrams were embedded.

I believe that Della Porta's diagrams and the relevant rules can be seen as having the epistemic role of "paper tools" as described by Ursula Klein for the case of the Berzelian formulas of nineteenth century chemistry. In her monograph "Experiences, models, paper tools. Cultures of organic chemistry in the nineteenth century" (2003) Klein has convincigly argued that the symbolic representation of chemical processes taking place in the laboratory was a key epistemic step in allowing researchers to tentatively manipulate those processes and the substances involved beyond the limits of laboratory practices, and thus contributing to knowledge construction in the same way as laboratory apparatuses did:

> Although a paper tool is a metaphor, I claim that on many levels paper tools are fully comparable to physical laboratory tools and instruments and that both kinds of tools contribute to the creation of reference and meaning, or representation in that sense.[14]

Paper tools could also contribute to the emergence of new scientific notions. Thus, the manipulations made possible by Berzelian formulas had "a suggestive power" that led a researcher "to go a step further than would have been necessary to construct his intended model. In an unintended way, this additional step led to the new concept of [chemical] substitution".[15] Similarly, Della Porta's diagrams led to the emergence of a notion of "point of inversion" for lenses, an essential step for better understanding both vision and the telescope.

Diagrams were paper tools with which Della Porta set out to explore, classify and "mathematize" optical experiences. Some recent studies on uses of diagrams in natural philosophy and scientific research have brought to light similar dynamics.[16] In his analysis of Feynman diagrams in high energy physics David Kaiser explicitly characterized them as "paper tools".[17] He showed how, depending on who was drawing and manipulating them, Feynamn diagrams could function as a rigorous formalism for physical-mathematical computations, or as a semi-quantitative, non-rigorous notation to grasp some properties of physical-mathematical models with which no rigorous computations were possible. In that case, the diagrams provided a plausible, but non-compelling demonstration of certain results, and they also led to the emergence of new physical notions, just like Della Porta's optical diagrams did. Important examples of the epistemic role of diagrams in the early seventeenth century include mechanical diagrams by Galileo and other authors, as well as the line-drawings employed in William Gilbert's treatise "De magnete" (1600), which Laura Georgescu convincingly argued helped "make the move" from experiments to concepts and theories and were "tools in Gilbert's reasoning about magnetic phenomena and magnetic motions".[18]

---

[14] Klein 2003, 245–46.

[15] Klein 2003, 188.

[16] Georgescu 2014, Heeffer 2017, Kaiser 2005, Mumma and Panza 2011, Wright 2014.

[17] Kaiser 2005.

[18] Bertoloni Meli 2006, Georgescu 2014, quotes from Georgescu 2014, 18, 24.

4.3 Optical Diagrams in *De refractione*: Tradition and Innovation

In his treatise *De refractione* (1593) Della Porta formulated rules for refraction in the glass sphere by "thinking" that artefact in terms of the spherical mirror, to whose optical understanding he had largely contributed.[19] Later in the same work, he used his rules for refraction in the glass sphere as a basis for treating lenses. To understand the challenges faced by Della Porta when studying lenses, it is therefore important to be aware of how problematic the models he was building upon were. I am referring here not only to the glass sphere, but also to the spherical convex mirror which, despite the great progress made in studying those devices, still presented fundamental conceptual problems, especially in the definition of its "point of inversion". Today we conceive image production in convex spherical mirrors in terms of its "focal point", which is located halfway between the centre of curvature of the mirror and the vertex of its surface. In a first approximation, all rays hitting the surface parallel to the principal axis of the mirror can be regarded as converging to a single point (focal point), but actually they are concentrated not on a point, but rather along a short segment (spherical aberration). Moreover, today's geometrical representation of the point of inversion as the focal point does not relate in a straightforward way to the apparent inversion of the image of an object which is experienced when moving the object or the eye (or both) with respect to the mirror. The empirical difficulty of locating a visual point of inversion is discussed in detail in the contribution by Yaakov Zik and Giora Hon to the present volume, where they show how Della Porta was aware that neither a geometrical "burning point" nor a visual "point of inversion" (i.e. the place at which image inversion seems to occur) of a convex spherical mirror were uniquely definable.[20]

Practitioners of optics from the late sixteenth century were all in some degree aware of the difficulty of finding a precise location for the focal point and for the point of inversion of a convex mirror, but they also knew that their approximate location was similar, i.e. somewhere between the centre of curvature of the mirror and the middle of point between that centre and the vertex of the surface. Beyond this vague similarity, though, there was at the time no reason to identify the two points, which were linked to two very different effects. Nevertheless, the two locations were often identified and placed at times at the centre of curvature, at times at the focal point. This situation may appear confused if one looks at it from the constrained perspective of today's geometrical optics, where a "point" is expected to correspond to a precise, point-like position defined in a theory. However, in the early modern period the relationship between geometrical principles and observed phenomena was much more fluid, as also Bertoloni Meli (2006) noted in his study on mechanics. Early modern practitioners of optics were primarily interested in finding a general representation of mirrors (and lenses) which fit both their burning

[19] Borrelli 2014, Smith 2015, 344–349, Zik and Hon 2012.

[20] Zik and Hon 2017.

capabilities and their perceived visual effects, and the techniques of lens-making of the time hardly allowed for a great precision of measurement. From this point of view, the "point of inversion" was a constitutively ambiguous concept and, as we shall see presently, it remained such in Della Porta's treatment of biconvex lenses. Although Della Porta was well aware of the difference between the geometric and the visual definition, he almost always employed the same term "point of inversion" to refer to what we today regard as different notions, suggesting that he was trying to identify the two concepts more closely.

The point of inversion was a relatively recent notion in the early modern period. The cathetus line, on the other hand, had a tradition reaching back into Antiquity, and its primary function was to help locate in three-dimensional space the apparent position of an image seen by the eye through a mirror or a refracting surface. The earliest and most authoritative definition of the cathetus was given for the case of reflection on a plane surface by Euclid: the cathetus was a line perpendicular to the surface of reflection and passing through the object being reflected.[21] To locate the point in space at which a given observer would see the image of an object, one first traced the reflected ray propagating from the object to the eye, and then drew the cathetus line relevant to the object: the image would be seen at the point of intersection between the cathetus and the extension of the reflected ray. These rules allowed a plausible prediction of where an observer would have the impression that the image of the object is located: behind the mirror at the same distance from it as the real object. The cathetus rule for plane reflection could be easily extended to reflection in a spherical mirror, where the perpendicular to the reflecting surface coincides with a radius of the spherical surface. The cathetus line for spherical reflection was regularly employed around the middle of the sixteenth century, and Della Porta would later appropriate it to treat spherical refraction.

In DR Della Porta gave the earliest detailed analysis of spherical mirrors found in print, and used it as a basis for formulating a rule of refraction for the glass sphere which led to qualitatively, and possibly also quantitatively plausible results. The mediation between the mirror and the sphere took the form of geometrical diagrams in which a circle drawn was seen first as representing the mirror, and then the sphere, with the same ray-tracing procedure being transferred from the first case to the second one. Characteristically for him, Della Porta offered no abstract justification for this analogy, but rather supported it by showing how a series of optical experiences with a refracting sphere could in this way be diagrammatically represented and qualitatively explained by adapting the rules of ray-tracing for the convex mirror, and in particular the cathetus rule. Della Porta defined the cathetus for the glass sphere as the radius of the sphere passing through the object, in a straightforward adaptation of the mirror's definition. He then used the rule for refraction in the sphere to trace the refracted ray and located the refracted image at the intersection of ray and cathetus. This procedure was very effective, since comparison with

[21] Smith 2015, 59–62.

experience showed that in most cases the image was indeed seen at that place.[22] Given these results, we can understand why Della Porta, when he went on to tackle refraction in spherical lenses, tried to adapt to the needs of the new artefact not only the rule of refraction and the point of inversion, but also the cathetus rule. Far from remaining attached to an obsolete principle, he was here following upon a very succesful practical approach.

What about the "point of inversion" for the sphere? I have already discussed how ambiguous the notion of "point of inversion" for the mirror was, and it is interesting to note that Della Porta in DR used very little both the term and the notion. He never spoke of the "point of inversion" of the convex mirror (DR I 1), but only employed that expression when discussing refractive artefacts. When speaking of the sphere, he wrote "point of inversion" only to indicate the position in whose vicinity image inversion occurred (DR II 16, 18), while avoiding the term when referring to the points where the sphere concentrated rays and could light fires (DR II 22, 23). Moreover, in DR the point of inversion played only a marginal role in image construction in the glass sphere. As we shall see, the same was valid for lenses.

4.4 Biconvex Lenses in Chapter VIII of *De refractione* (DR VIII)

Della Porta discussed lenses in chapter VIII of *De refractione* (DR VIII), and this part of the treatise seems more a report on work in progress than the presentation of final results. The general idea was to extend to lenses the methods used for the sphere, and indeed proposition DR VIII 2 was entitled "Finding the point of inversion in convex lenses"[23] and Della Porta explained that refraction in a spherical convex lens could be treated by applying the rule valid for the glass sphere. This was a straightforward solution, and both in the treatise and in the manuscript Della Porta followed it, although in his diagrams he rarely bothered to actually construct the refracted rays according to the rigorous procedures, but simply drew them in a way which usually, but not always, seemed qualitatively plausible. The point of inversion for planoconvex lenses was defined in DR VIII 2 as "the point where refracted rays meet"[24] and located by using ray construction. However, Della Porta said nothing on how to locate the point in which rays converged for biconvex lenses.

Having provided a rule for tracing refracted rays in biconvex lenses, in principle all Della Porta needed to do now was to define a cathetus for the lens and then locate

[22] Della Porta's cathetus line for the glass sphere had the same path as a ray emanating from the object and passing through the centre of the sphere, which therefore remains unrefracted. The refracted ray, as already noted, was drawn according to a qualitatively plausible rule, so that the intersection of ray and cathetus actually provided a good approximation of where an image would be seen, as I have discussed in detail in (Borrelli 2014).

[23] "in specillis convexis inversionis punctum invenire" (Della Porta 1593, 175–176).

[24] "punctum concursus refractarum linearum" (Della Porta 1593, 175).

the image as usual at the intersection of ray and cathetus. However, the adaptation of the cahetus rule to biconvex lenses was not straightforward, since those lenses have two spherical surfaces with no common centre, and so no line could be drawn perpendicular to both of them. One was therefore forced to choose a cathetus line perpendicular to only one surface, and so in principle the cathetus could be equally well drawn through the one or the other of the two centres of curvature. Before going on to see how Della Porta solved this problem in DR, I shall introduce a convention for diagramatically representing lenses which the Neapolitan scholar employed throughout that work and in all his subsequent writings on lenses: a lens was alway represented as lying horizontally, with the object above it and the eye below it (Fig. 4.1a, where AB is the object and E the eye).

In this way, one could uniquely identify the two surfaces of the lens as upper ("superior") and "lower" ("inferior") one. The circles corresponding to the two surfaces were also indicated as "upper" and "lower", although in the diagram the lower circle was above the upper one. The convention allowed to unambiguously describe the positions of eye and object as being either "above" or "below" the point of inversion of the lens' surface facing them. Thus, if the eye was described as lying "above" the point of inversion, this meant that it was nearer to the lens than that point; on the other hand, if the object was "above" the point of inversion, it was further away from the lens that the point itself. The great attention paid by Della Porta to these details should not be underestimated, as it proves how aware he was of the importance of coordinating text and diagrams in multi-medial message to the reader.

Expressed in terms of the above convention, the cathetus line chosen by Della Porta in DR was the line passing through both the object and the centre of the lower circle, i.e. of the circle of curvature of the side of the lens facing the eye. Della Porta

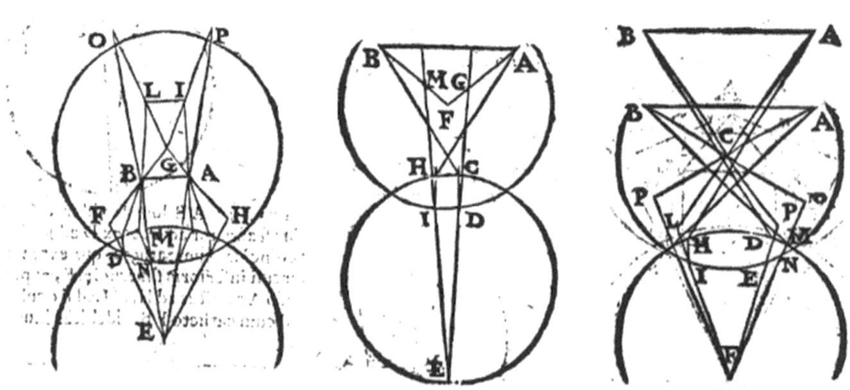

1a (DR VIII 7) 1b (DR VIII 8) 1c (DR VIII 9)

Fig. 4.1 **(a–c)** *De refractione*, Chapter VIII, proposition 7, 8, 9

offered no explanation for this choice, but proceeded to state and demonstrate with diagrams a series of propositions on image formation for biconvex lenses:

- If the eye (DR VIII 3) or the object (DR VIII 4) or both (DR VIII 5) are located at the point of inversion, the image of the object will fill the whole lens;
- If the object is near the lens, whether the eye is near or far, the object will appear upright (DR VIII 6);
- If the eye is near the lens, whether the object is near or far, the object will appear upright (DR VIII 7);
- If both eye and object are far from the lens, the object will appear inverted and nearer (DR VIII 8), and the further away they are, the smaller the object will appear (DR VIII 9);
- If the object is below the centre of curvature of the lens, the nearer it is to the eye, the larger it will appear (DR VIII 10).[25]

The final propositions of the chapter (DR VIII 11 to 16) dealt with oblique vision, biconcave lenses and burning mirrors and I shall not discuss them here. All propositions were empirically valid, if at times vaguely fomulated. Let us consider the role of the point of inversion: after having been defined geometrically in DR VIII 2, it only appeared in the first three propositions, when the object and/or the eye were placed exactly at the point of inversion. In this way, the point of inversion was visually characterized as the position at which the image filled the lens. The configuration eye-lense-object resulting in an upright or inverted image was instead only specified by saying that the eye or object were "far" or "near" the lense. This is perhaps not surprising, since, as we saw above, Della Porta had not given any precise location for the point of inversion of a lens and so he could not connect that notion to the visual experience of image inversion. In conclusion, although Della Porta gave the point of inversion a prominent position in the structure of the chapter, he did not, or probably could not, actually embed it in his ray tracing rules beyond the rather generic statements of DR VIII 3–5.

I will now take a closer look at propositions DR VIII 7, 8 and 9, which discussed the occurrence (or not) of image inversion. Proposition DR VIII 7 stated that, when the eye is near the lens, the object will be seen upright no matter whether it is near or far from the lens (Fig. 4.1a). Let us consider the image of object LI: the ray from L enters the lens in M, exits it in N and goes on to the eye E. The ray EN is then extended back to O, where it meets the cathetus GLO. The same construction applies to I, and the image is upright. Although this construction is plausible, we may note that the paths taken by the rays from I and L to the eye are rather peculiar, since they look as if the rays were being refracted beyond the perpendicular when they enter the lens. As we shall see later on, Della Porta, too, eventually started to regard this construction as problematic.

[25] In the title of the proposition the "centre" is indicated as "centre of the eye" ("centrum oculis", Della Porta 1593, 181), but in the description it is called "centrum circuli" and the diagram leaves no doubt that the title was a misprint.

In proposition DR VIII 8, Della Porta presented a construction (Fig. 4.1b) showing how a far-away eye (E) would see a far-away object (BA) inverted and nearer. The text of the proposition explained that a ray from B reaches the lens in C, is refracted to D, and from there is refracted again to reach the eye E. The same happens to a ray from A, which passes through H and I to reach the eye E. The lines ED and EI are then extended back to meet the two catheti starting from the centre (F) of the lower circle and passing respectively through A and B. The cathetus passing trough A crosses the extension of the ray emanating from B at point G, where the image of B will be seen. According to an analogous construct, the image of A will be seen at point M and the object appears inverted. The inversion was therefore due to the fact that the refracted rays crossed each other before reaching the eye, so that the eye received from the right side the ray originally coming from the left and vice versa. This interpretation corresponds to the traditional view, often endorsed by Della Porta, that the (refracted) ray carries the "image" ("species") of the object, while the cathetus only determines the image's position. However, relating image inversion to ray crossing in this case begs the question, why the crossing occurs in the first place. The diagram somehow suggests that, if the rays did not cross, they would not reach the eye, but no attempt at a demonstration of this fact was made. While in the case of reflection in the convex mirror it was easy to show (or even rigorously prove) that in certain cases the rays could only reach the eye if they crossed each other, here the situation was more difficult. Della Porta's rule for refraction in spherical surfaces was too complex to allow him to give a rigorous proof that in some cases the rays had to cross to reach the eye. Indeed, it may remain open to discussion whether or not such a proof is in principle possible. Therefore, the objection could be raised that perhaps the rays from A and B could have reached the eye in E even if they had not crossed each other, so that image inversion would not have occurred. We shall come back to this issue later on. A further point worth noting is that in this diagram, other than in the usual perspectivist constructions, the ray carrying the image of A is cut by the cathetus drawn from B and vice versa: here, too, a reversal took place.

Proposition DR VIII 9 stated that an inverted image would appear larger or smaller depending on whether the eye and the object approach or move away from the lens. The diagram in Fig. 4.1c showed how to construct the image of an object AB far from (image LM) or near to (image OP) the lens. As in DR VIII 8, the image was inverted because the rays crossed, but this time the cathetus lines did not cross and the cathetus from A [B] intersected the ray emanating from A [B]. Although it would have been possible to let the cathetus lines cross each other, the construction would have probably resulted in an empirically incorrect prediction, with the image becoming larger if the object moves away from the lens. Here we have a first example of how Della Porta was using the cathetus as a flexible tool to construct diagrams in agreement with experience, albeit at the expense of theoretical coherence. Della Porta's diagrammatic constructions may be open to many questions, but they fulfilled their goal, showing how given principles of lens refraction could account for observed phenomena. However, on the whole the net results for lenses in the *De refractione* were not very impressive: the "point of inversion" of the lens was defined as the focal point, but was not located and was also not related to the phenomenon of image inversion. More in general, no justification was offered why under certain circumstances ray crossing occurred, resulting in an inverted image. Della Porta would try to improve on these points in the following years, as the draft of the treatise *De telescopio* shows.

4.5 The Draft Treatise *De telescopio*

In 1940 an authograph manuscript by Della Porta was found in the archive of the Accademia dei Lincei. It was edited in 1962 by Maria Amalia Naldoni with an introduction by Vasco Ronchi and is now kept in the Library of the Accademia dei Lincei in Rome.[26] The manuscript is dated to 1610 according to the Library and ca. 1612–1615 according to its editors, and its leaves were bound together in a sequence which does not correspond to the order in which they were written. Its first part contains an autograph of Della Porta's treatise *Elementorum curvilineorum libri tres*, while the rest consists of various drafts of the chapters of a treatise *De telescopio* (DT) which, according to Ronchi's reconstruction, Della Porta had started writing around 1612 in context of his priority dispute regarding the telescope, but never had time to finish.[27] The contents of the DT are an extension and revision of the material presented in DR, and were discussed by Naldoni and, as far as the chapter on the telescope is concerned, by Ronchi.[28] Naldoni dealt in most detail with the chapters on refraction in the glass sphere and praised the way in which Della Porta had progressed in his experimental work, of which she offered detailed descriptions, yet she joined Ronchi in his disparaging judgement of the Neapolitan scholar's attempts to derive laws of refraction for sphere and lenses. Instead, I will argue that Della Porta's treatment of biconvex lenses diplayed a high degree of methodological coherence and represented what would become a heuristic template for later experimental practices in optics and other disciplines. Apart from Ronchi and Naldoni, the draft treatise DT has been discussed by Fumikazo Saito in his monograph on the role of the telescope in Della Porta's natural magic.[29] In that volume Saito argues that Della Porta employed geometrical constructions as a means to connect optical instruments with the object of investigation, and claims that the Neapolitan scholar was not interested in understanding the functioning of lenses, but only in showing, through constructions, that a match between instrumental practice and phenomena was possible.[30] From this point of view, it was not so important whether Della Porta's constructions followed some system, provided that they appeared plausible, and so Saito does not reconstruct Della Porta's geometrical reasoning, and seems to

[26] Della Porta [1610–1615], Della Porta 1962.

[27] Naldoni 1962, Ronchi 1962.

[28] Naldoni 1946, 1962, Ronchi 1962.

[29] Saito 2011, see also Saito 2008.

[30] Saito 2011, 219–224.

follow Dupré in attributing to the diagrams a purely rhetorical functions. While agreeing with Saito's key thesis on the telescope, I will argue that there is more to the diagrams of DT than meets the eye and that it is worth looking in detail at the way in which Della Porta developed his strategy for constructing them.

The titles of the draft chapters of DT are:

Chapter 1: On light rays passing through transparent bodies (extant in three versions)
Chapter 2: On the crystal sphere (extant in four versions)
Chapter 3: On biconvex lenses (extant in three versions)
Chapter 4: On biconcave lenses (extant in two versions)
Chapter 5: On the telescope (extant in one version).

The manuscript also contains alternative chapters "four" and "five" on planoconvex and planoconcave lenses (i.e. semilenses). As noted by Naldoni, the alternative versions often differ very much from each other, and in the following pages I will discuss only the three versions of the chapter on biconvex lenses (chapter 3). I will label these three versions DT 3A (f.111–116), 3B (f. 132–139) and 3C (f 142–145). Naldoni stated that version 3A seemed to her the most recent one and that it appeared more ordered than the other two. She suggested that 3C might be the oldest one, but I will argue on the ground of the contents that the chronological sequence is actually 3B-3C-3A.

All versions of chapter 3 take the form of series of propositions composed of text and diagrams. However, a closer reading shows that, especially in version 3B and 3C, this structure was not the final result of an investigation, but rather a report of a process of ongoing reflection. As we shall see, some propositions contradicted each other, or turned out not to be compatible with experience, prompting Della Porta to revise them in later versions. The scheme of all draft chapters was the same one: Della Porta started by assuming some rules for writing down diagrams, deduced from them a series of implications on what could or not be seen in the lenses, and eventually checked how far these results fit observation. More often than not, he had to go back and tentatively re-adjust the procedures to trace the refracted ray and draw the cathetus. While the rules for constructing rays refracted by a spherical glass surface remained stable, the cathetus line changed time and again both in its definition and in its function in the diagram. No a priori justification of the whole trial and error procedure was offered and, while Della Porta has usually been criticized for not having a theoretical grasp of what he was doing, one can see his modus operandi here as perfectly fitting what would later on become the tradition of experimental philosophy, a fact which has been so far largely ignored by historiography. About the different definitions of the cathetus lines in the draft chaperts of DT Naldoni wrote:

> Each version is a new attempt. Each time Della Porta starts from a scratch, completely modifying, or even turning upside down the principles on which his theory is based. It is no use searching for some progress in all this: it is a blind groping in search of some system

with which to validate experiences, but without finding a way out. And this is natural, as we are here on the wrong track.[31]

Naldoni's remarks are an example of the biased perspective from which she, like Ronchi and many later authors up to and including David Lindberg, judged Della Porta: they tried to grasp his reflections in terms of a "theory" and compared them with modern theories of light and vision, without paying any attention to Della Porta's methodology and to his strategies of bottom-up conceptualisazion.[32] Ever since the 1980s, though, historians and philosophers of modern and early modern experimental practices have shown how such practices constituted an epistemically independent element of the investigation of nature, and a key component of modern science. Accordingly, I will endeavour to show how Della Porta's changes of approach in the different versions of DT were part of a heuristic strategy at whose centre stood a close mutual feedback between experience and conceptualization. In fact, I shall argue that Della Porta's way of investigating lenses displays precisely those features of experimental practices which recent studies have highlighted: an epistemic dynamics largely independent of broad theoretical reflections and closely linked to the material, performative and representational aspects of experimental activity.[33]

4.6 Biconvex Lenses in *De telescopio* Draft Chapter 3B (DT 3B)

One reason why I believe that version DT 3B is the oldest among the extant drafts of chapter DT 3 is that it employs the same definition of the cathetus for biconvex lenses that we found in DR VIII. However, both the classification of the optical effects of the lens and the diagrams often differ from those of the published treatise. One feature remains constant, though: in DT Della Porta adheres to the same conventions for diagrammatic representation of lens refraction employed in DR VIII: the object is always above the horizontally placed lens and the eye below it, and the two spherical surfaces of the lens are accordingly labelled as "upper" ("superior") and "lower" ("inferior") one. This convention is in DT even more important than in DR, since the two circles and their centres become important parameters for describing the workings of lenses.

Chapter DT 3B contains various propositions similar to those in DR VIII discussing image inversion and magnification, but now the point of inversion plays a

[31] "Ogni copia e' un tentativo diverso. Ogni volta il Porta ricomincia da capo, modificando completamente, anche addirittura capovolgendo i punti su cui basa la sua teoria. Ed è inutile cercare in tutto questo una linea di progresso: è un brancolare cieco in cerca di un qualche sistema con cui poter convalidare le esperienze, ma senza trovare una via di uscita. Ed è naturale perchè siamo fuori strada" (Naldoni 1946, 67).

[32] I have discussed these issue, and particularly Lindberg's bias in (Borrelli 2014).

[33] Galison 1997, Klein 2003, Pickering 1995.

more important role as a parameter for determining when these effects occur. We recall that in DR VIII Della Porta had offered diagrams showing how an object far from the lens was seen inverted, and one near it was not, but had not explained why these specific ray configurations obtained in those two cases, nor had he linked them to the "point of inversion". This is what he endeavoured to do now. Proposition DT 3B 1 discussed how to find the point of inversion in biconvex lenses, but did not introduce the geometrical focus of the lens. Instead, Della Porta started by describing in words and diagrams two possible constructions of how an object placed "above the point of inversion" could be seen through the lens: inverted or upright. These constructions had the same structure as those appearing in DR VIII 9 and DR VIII 7.

In the first one (Fig. 4.2a) the rays from the object AB crossed each other before entering the lens in G and D, exited it in H and D, and finally reached the eye F; the catheti were drawn from A and B through the centre C of the lower circle, and did not mutually cross. The object AB was seen inverted and smaller at the intersection KI of rays and catheti. After having described the first construction, Della Porta went on to say that, if someone should object and claim that the rays from A and B did not cross each other, but travelled straight down to the lens and from there to the eye, then the opposite visual effect would obtain: the image would appear upright and magnified. He demonstrated this in the diagram in Fig. 4.2b, in which the rays followed the paths ACD and BFG, to meet at the (unlabelled) eye. The two catheti drawn from the centre of the lower circle K through A and B intersected the extensions of the rays in H and I and the image was upright and magnified. Both constructions appear qualitatively plausible, and Della Porta did not say which

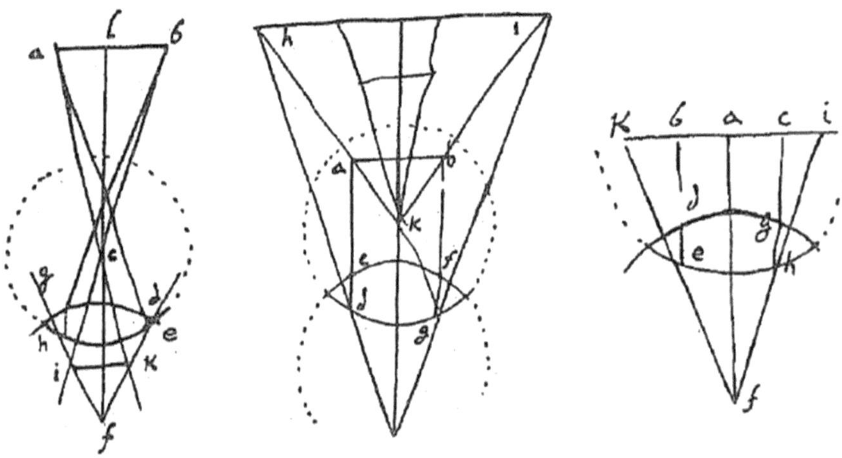

2a (DT 3B 1a) 2b (DT 3B 1b) 2c (DT 3B 1c)

Fig. 4.2 (**a–c**) *De telescopio*, chapter 3B, proposition1

diagram would apply and when, but rather went on to describe a third case in which the object was located exactly at the centre of the lower circle. Its image was also located there and appeared magnified (Fig. 4.2c). He then concluded: "From these experiences we are led to locate the point of inversion along the upper semi-diameter of the lower circle".[34] Thus, Della Porta here characterized the "point of inversion" as the point in whose vicinity image inversion occurred, and used diagrammatic constructions to locate it. No mention of the geometrical focus was made, so that the point of inversion was defined exclusively in visual terms. The argument for the position was not compelling, and indeed quite vague: it was rather an attempt to see whether the rules for image construction in lenses introduced in DR VIII could be employed to determine the location of a visually defined point of inversion. To this aim, the fact that the cathetus was drawn from the centre of the lower circle was crucial, since the relevant diagrams made visually plausible why, when the object was also placed there, the image could only be seen upright.

Having defined and located the point of inversion of the lens Della Porta went on to reprise in DT 3B 4 one of the two alternative constructions proposed in proposition DT 3B 1: the one without ray crossing and with an upright image (Fig. 4.2b). Proposition DT 3B 4 stated that an object located above the point of inversion would always be seen upright. This claim clearly contradicts experience, and Della Porta indeed later wrote near it: "wrong" ("falsa"). Propositions DT 3B 5 and 6 related to the magnification of objects lying below or at the point of inversion, and here the same diagrams as in DR VIII appeared. Proposition DT 3B 7 and 8 stated that an object above the centre of the lower circle would be seen on or near the surface of the lens, a statement not contained in DR VIII which might indicate that Della Porta was including new empirical material in his study.

Proposition DT 3B 9 addressed again the difficult issue of image inversion. Contradicting "wrong" proposition DT 3B 4, it stated that an object located above the point of inversion would never be seen upright, and demonstrated this statement using the series of diagrams shown in Fig. 4.3.

The first three of these diagrams depicted three possible paths of rays propagating from the extremity of an object (A, I, R) to reach the eye (E, N, A): one ray parallel to the diameter of the lower circle, and two rays inclined with respect to it towards the left or the right. In all three cases, Della Porta claimed, the cathetus drawn from point A/I/R through the centre of the lower circle (F/O/X) would never intersect the extension of the refracted ray on the same side of the vertical diameter as A/I/R. Accordingly, no upright image could ever be formed. The fourth construction purported to show how the cathetus drawn from A would intersect the (extension of the) refracted ray from A if the ray crossed to the other side before entering the lens, thus giving rise to an inverted image. What were the goal and the validity of this proposition? As to the goal, Della Porta was here trying to demonstrate that the image of an object far away from the lens appears inverted by showing that ray

[34] "Ex his experientiis ducti indicandum inversionis punctum est superius inferioris circuli per semidiametrum existere" (Della Porta 1962, 113). Naldoni read "superior" instead of superius, but I believe that both the content and the writing in the manuscript support my reading.

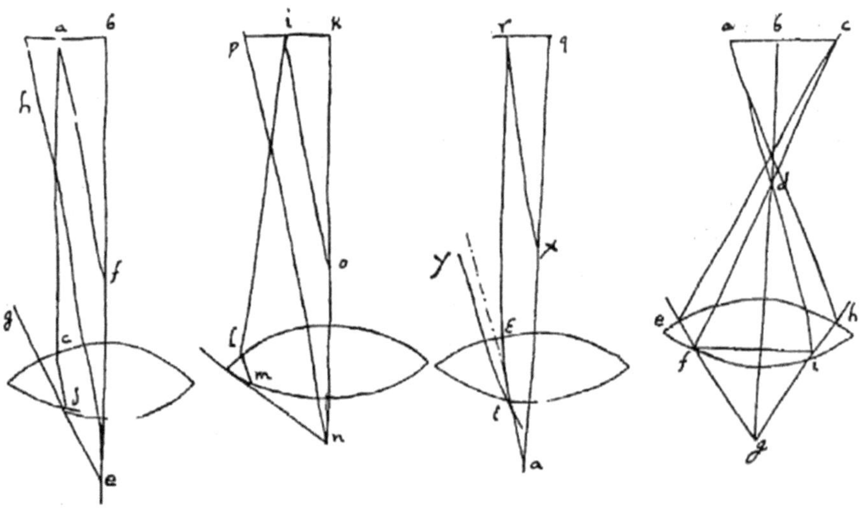

Fig. 4.3 *De telescopio*, chapter 3B, proposition 9

crossing had to occur for the ray to meet the cathetus. How far was the proof valid? Although the constructions were carefully drawn and qualitatively plausible, the paths of the rays were clearly not rigorously deduced from the rules of refraction in the sphere. Indeed, Della Porta did not even attempt such a deduction, but rather presented a simplified semi-quantitative argument, such as those employed often in bottom-up theorizing and exploratory experimentation.[35] The diagrams had here indeed a "rhetoric" function, but not in the sense that they were icons of geometrical exactness, but rather that they tried to convey to the reader the gist of a hypothetical rigorous proof which could not be performed. Finally, it must be noted that there was still a catch in proposition DT 3B 9: an object far away from the lense is still seen upright if the eye is very near the lense, as stated by Della Porta already in DR (DR VIII 7). Significantly, that statement was not repeated in chapter DT 3B, suggesting that Della Porta had decided to discard it because he was aware of the contradiction and regarded proposition DT 3B 9 as more plausible. We shall see later on that he eventually changed his mind.

Propositions DT 3B 10 and 11 discussed again magnification, and were largely a reprisal of propositions DT 3B 5 and 6. Finally, proposition DT 3B 12 summarized the results: "In biconvex lenses, when the object is beyond the point of inversion, its image is smaller and inverted, [if the object is] on the circumference of the lower circle [the image will be] equal and inverted, [if the object is] below the centre [of

[35] Borrelli 2015, Steinle 2016.

the lower circle, the image will be] upright and larger".[36] The corresponding diagram combined the constructions presented in propositions DT 3B 9, 10 and 11. The remaining part of DT 3B cannot be discussed here, and I only wish to note that in proposition DT 3B 14 Della Porta stated that, when an object is beyond the point of inversion, various images of it could be seen, among them one inverted and "hanging in the air": this was the first time that Della Porta mentioned such an image as produced by lenses, although in DR he had described how the glass sphere created images "hanging in the air".

Summing up the discussion of DT 3B, I believe it can be characterized as a first attempt to flesh out the account of lenses given in DR VIII, and in particular to offer a diagrammatic model of how the inversion effect seen through the lens could be related to a (somehow defined) "point of inversion". To that aim, Della Porta first provided a definition and a location for the point of inversion, choosing to characterized it in purely visual terms, without identifying it with the focal point of the lens. The location of the point of inversion so defined was argued on the basis of a diagram in which the definition of the cathetus as passing through the centre of the lower circle constituted the key element of the argument. Thus, the choice of the cathetus determined the point of inversion. The final result of the investigation went quite a few steps further than DR VIII, but still displayed a number of problems.

4.7 Biconvex Lenses in Draft Chapter DT 3C

In chapter DT 3B Della Porta had offered empirically plausible diagrammatic reconstructions of image inversion and magnification in biconvex lenses, and related them to the position of the object with respect to the point of inversion of the lens. This point had in turn been defined purely visually and located along the upper semidiameter of the lower circle on the basis of the choice made on how to draw the cathetus. However, proposition DT 3B 9, which stated that an object beyond the point of inversion always appears inverted, contradicted observations already known to Della Porta at the time of DR. This problem was addressed right at the beginning of DT 3C, where it was solved by modifying the definition of the cathetus line. Proposition DT 3C 1 had the title: "Investigating the centre of biconvex lenses"[37] and its main theme was "whether the cathetus line should be drawn from the centre of the lower circle or of the upper one", an issue which "should be tentatively approached on the basis of both reasoning and experience".[38] Della Porta's first move was to state that images in biconvex lenses are often seen upright and smaller,

[36] "Magnitudine in specillis utrinque convexis extra inversionis punctum constituta imago minor et inversa, in circumferentia inferioris circuli aequalis et conversa, infra centrum recta et maior" (Della Porta 1962, 122).

[37] "Centrum in specillis utrinque convexis investigare" (Della Porta 1962, 128).

[38] "Cathetus ex inferioris vel superioris circuli centro ducenda, prius discutendum est quod ratione et esperientia tentandum est" (Della Porta 1962, 128).

and then to show how this observation fit a diagrammatic construction in which the cathetus line was drawn from the centre of the upper circle, i.e. the opposite choice to the one made in DR and DT 3B.[39] He then expounded his arguments against the previous definition of the cathetus for the lens, and did so by using precisely the same constructions which had appeared in proposition DT 3B 9 (Fig. 4.3). As before, he argued that, by choosing a cathetus passing through the centre of the lower circle, the image of an object located beyond that centre would never be seen upright, but this time he concluded:

> The point A will be seen in I, and C in F [i.e. the image is inverted], whereas we know by experience that an object will always be seen upright if the eye is near the lens, and this is the opposite [of what the diagram says]. Therefore we should not consider [for the cathetus] the centre of the lower circle[40]

In other words, Della Porta had now realized that proposition DR VIII 7 was correct, while DT 3B 9 was not. He reacted by modifying the diagrammatic rules in their most flexible feature: the cathetus. Although we may today regard Della Porta's rules of image formation flawed for their reliance on the cathetus line, we must acknowledge that here he was deriving them according to a quite consistent and in principle efficient methodology: building a simplified model for geometrically conceptualizing experience, using this model to derive further predictions, and comparing them to experience. The absence of any kind of more abstract reflections on universal laws of light and vision was in this methodological context an asset rather than a drawback, and the notion of cathetus line had the function of a flexible paper tool to explore and connect optical experiences. For Della Porta the cathetus was a necessary tool, since optical images were defined through the intersection of two lines, and giving up the cathetus would have constituted a serious problem. Accordingly, having disposed of the "wrong" cathetus, Della Porta now faced the problem of finding the "right" one. How did he go about it? By trial and error, as was often the case in his writings. After having shown that the cathetus choice in DR and DT 3B was wrong, he did not go on to explicitly state how the right cathetus line was defined, but instead turned to the definition of the point of inversion. Let us follow him.

In DT 3C, Della Porta defined the point of inversion as the focal point, i.e. the region in which rays were concentrated by the lens. In DT 3C 2, he presented a construction of the focal point of the glass sphere, showing how the rays were not concentrated in a point, but rather along a segment around a point located at a distance from the sphere's surface equal to its radius. Because of a second refracting surface which further concentrated the rays, so Della Porta argued, a biconvex lens would concentrate rays more than the sphere, although it would still have no point-like focus. It is not clear where exactly Della Porta placed the focal segment of the

[39] As usual, the path of the refracted ray was derived in principle according to the refraction rule for the sphere, although here, too, the actual drawings were very approximate.

[40] "Nam A in I, C in F conspicabitur, ubi experientia cognoscimus rectam semper videri oculo propinquo, quod est contrarium non igitur superius centrum inferioris portionis attendatur" (Della Porta 1962, 129).

lens, partly because the hand-written passage is not well readable, but probably also because the author himself was not sure.

In the remaining of DT 3C, despite having stated that the lens striclty speaking had no geometrically exact point of inversion, Della Porta usually spoke of "the" point of inversion of a lens and drew it in diagrams sometimes at the lowest point of the upper circle, but more often slightly lower than that, as shown for example in Fig. 4.4 (DT 3C 7), where the point of inversion is F.[41]

Having characterized the point of inversion, in the following propositions Della Porta always drew the cathetus line passing through it, and on this basis proceeded to state, among other things, that an object near the lens would always be seen upright (DT 3C 3), that (otherwise) an eye below the point of inversion would always see the object inverted (DT 3C 6), and that an eye above the point of inversion would see the object always upright (DT 3C 7).[42]

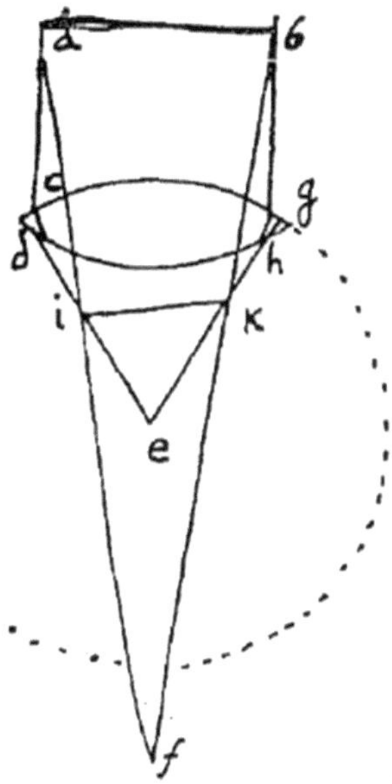

4 (DT 3C 7)

Fig. 4.4 *De telescopio*, chapter 3C, proposition 7

[41] Of course Della Porta knew that the lens had two symmetrical points of inversion, but in DT 3C he only used the one towards the eye, because he classified the effects according to whether the eye was above or below the point of inversion. In DT 3B, he had done the opposite, classifying all effects according to the position of the object.

[42] Naldoni's trascription of the title of DT 3C 6 states that the eye is above ("supra") the point of inversion (Della Porta 1962, 132), but text and image clearly refer to the case in which the eye is below it. A comparison to the manuscript shows that the word Naldoni read as "supra" is in fact "infra" (Della Porta [1610] f. 144r).

The location of the point of inversion and its coincidence with the centre for cathetus lines might seem an unmotivated, chaotic move, but it was a very important step in Della Porta's scheme. Although this definition was a clear breach of the principle that the cathetus line should be perpendicular to the refracting or reflecting surface, for Della Porta it was worth making this radical change since the new definition provided him with a means for diagrammatically explaining inverted images. Since the cathetus started from the point of inversion, if the eye was above that point, neither the catheti nor the refracted rays could cross, and so Della Porta could claim that, in this case, the image was always upright (DT 3C 7, Fig. 4.4); if the eye was below the point of inversion, on the other hand, the rays might or not cross, but the catheti necessarily did, and because of this the image was always inverted (DT 3C 6, Fig. 4.5).[43] In this way, image inversion was shown to follow necessarily from the positions of eye or object with respect to the point of inversion and from the rules for drawing the diagrams.

Let us take a closer look at this very innovative solution (Fig. 4.5): the ray from A was refracted in B, then in C, to reach the eye D (below the point of inversion E),

Fig. 4.5 *De telescopio*, chapter 3C, proposition 6

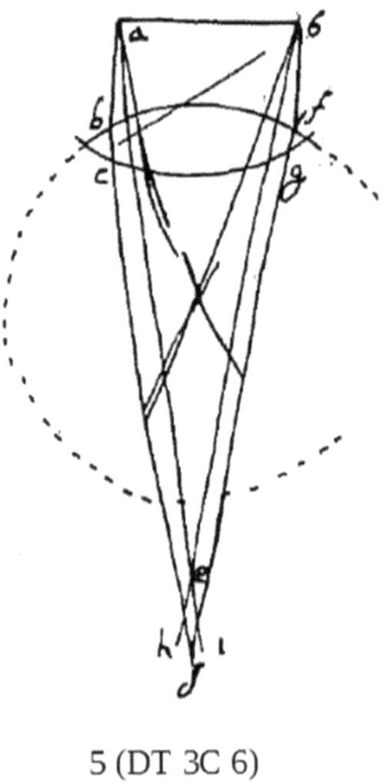

[43] Despite all efforts, there was here still a contradiction between DT 3C 3 and 6, since it is not clear how the cathetus crossing was avoided in DT 3C 3 when the eye is below the point of inversion. The relevant image is unhelpful and we may regard this as a point on which Della Porta needed to work further: he had now realized how the empirical situation was (i.e. the image is not always inverted if the object is above the point of inversion), and was looking for rules allowing him to make sense also of this result.

while the ray from B reached the eye by passing through F and G. The cathetus joining A to E intersected the refracted ray coming from B at point I, and the cathetus from B intersected the ray from A at point H. This process, according to Della Porta, resulted in image inversion: "so the eye will see on the right what is on the left and on the left what is on the right, and what is below will be seen above".[44] Image inversion therefore did not follow from a crossing of the rays, as had been the case in previous theories of refraction or reflection: the ray coming from the left (i.e. from A) reached the eye from the left, and the one from the right (i.e. from B) reached the eye from the right. However, the ray from A was cut by the cathetus from B, and at their intersection the image of B, and not of A, was seen. The opposite obtained on the other side, where A was seen at the intersection of the ray coming from B and of the cathetus coming from A. This solution was radically innovative, as it meant that the cathetus line was now being assimilated to the ray: it was now the cathetus, and not the ray, which determined whether the image of A or B would be seen at their intersection, and therefore in a sense the cathetus now "carried the image" of the point. Della Porta was here using the cathetus, the ray, and the point of inversion as flexible paper tools to formulate rules for constructing diagrams which fit experience.

4.8 Draft Chapter *De telescopio* 3A: Diagrams Between Experience and Geometry

As Naldoni noted, DT 3A is by far the most attentively written of the three versions of the chapter on biconvex lenses in DT, and I believe that it was the only one which Della Porta consciously wrote to introduce the subject of refraction in lenses in a pedagogically effective way. Therefore, I will offer a somehow more extensive treatment of it, analyzing not only Della Porta's results, but also the way in which he was presenting them to his readers, in particular as far as the role of diagrams was concerned. Proposition DT 3A 1 was devoted to defining what lenses and semi-lenses are and to finding their point of inversion, which was here defined as the place were rays are focussed. Della Porta summarized the procedure for identifying the point(s) where a glass sphere concentrates parallel rays, showing how these points were spread along the segment OH in Fig. 4.6a. He then argued that similar points for the biconvex lens would be further concentrated in the segment HI near the centre of the (upper and lower) circle (Fig. 4.6b).

As had been the case in DT 3C, although he had shown that biconvex lenses had no point-like geometrical focus, in the rest of the chapter he spoke of "the" point of inversion of each face of the lens and located it at the centre of the relevant circle. Why had the centre of curvature of the lens now become the point of inversion? No geometrical construction was offered to justify this choice but, later on in the chapter, the centre of the upper circle will be chosen as the point from which to draw the catheti. This choice would eventually allow Della Porta to employ in DT 3A the

44 "unde oculus quae dextra videbit sinistra et quae sinistra dextra et quae infra erunt supra spectabuntur" (Della Porta 1962, 132).

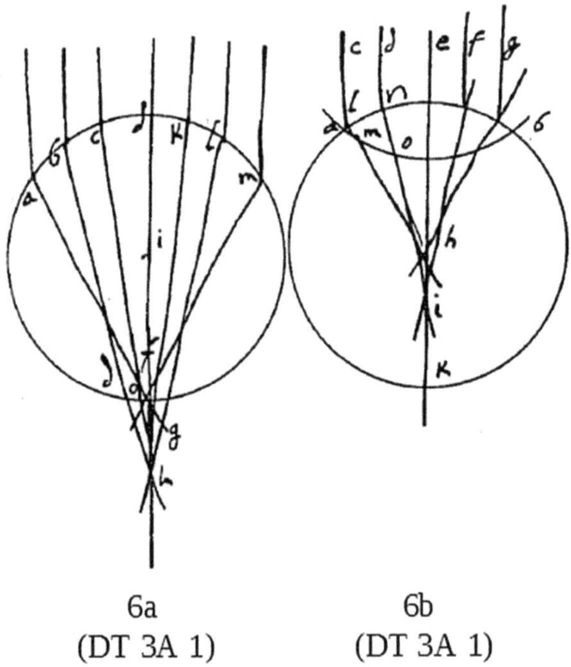

Fig. 4.6 (**a**, **b**) *De telescopio*, chapter 3A, proposition 1

same arguments used in DT 3C for relating image inversion and point of inversion, and so it may be assumed that this goal motivated the choice of latter's location. In other words, Della Porta wished to draw the cathetus through the point of inversion as he had done in DT 3C, but at the same time he wanted to define the cathetus as perpendicular to the refracting surface. Therefore, he (plausibly) argued that the point of inversion was located at the centre of curvature of the lens' face. However, Della Porta made no explicit statements about the reasons behind his choice, and simply drew the point of inversion at the centre of curvature. Let us now look more closely at his argument.

Having introduced the geometrical properties of biconvex lenses and their approximate geometrical focus, Della Porta now turned to the visual effects that they could produce. Proposition DT 3A 2 discussed "by means of which experiences the workings of lenses are to be investigated".[45] Della Porta explained that he would describe what one sees in lenses ("specillorum apparentiis"), and also represent what had been seen in appropriate drawings, with the aim of using them to discuss the way in which the various images change.[46] There followed first a verbal

[45] "Quibus experientiis specillorum operationes indagandea sint" (Della Porta 1962, 102).

[46] "Nunc quoque quomodo indaganda sunt inversionis puncta ex specillorum apparentiis, graphice appositeque inspectorum facies describamus ut ex horum disquisitione discriminum variantium figurarum aucupemur" (Della Porta 1962, 102).

description of how to manipulate a biconves lens, and what to expect to see in doing so, and then a second description of the same procedure, but this time combining words with a diagram geometrizing the experience. In the end, Della Porta noted that one should not regard the whole passage as unnecessary, as one must not be weary of repeating and going slowly, because the ease of learning depends on the way in which the discussion is made.[47] The first, purely verbal description is the following:

> One must take the lens in the hand and first of all put it in front of one's eye. When the object is located far away, it will be seen as it is, upright, clear and only a bit larger. Then let the lens move away slowly from the eye and this movement will let the object appear a bit larger, and continuing to move gradually in this way, the object will become so bloated, that it will fill almost all of the lens. After that, the image melts, becomes distorted and transformed, and the lens will be filled by darkness, as all of the refracted rays are made to converge to the same centre of inversion. Then, moving the lens further away with the hands so that it will lie midway between the eye and the object seen, the darkness will fade away and the object will appear inverted, shrunk within a small sphere and somehow clearer. And stretching further out the hand the object will grow larger, until it once again occupies all of the lens. Then the image is pushed out and disappears and darkness flows together and concentrates in the point of inversion, as only in the middle it is possible for the image to pass from upright to inverted. At the end, when the lens moves further, it arrives near the object, which will be perceived clear, equal, upright as it was when the lens was near the eye.[48]

In this text the qualitative, subjective experience of looking through a biconvex lens goes through a first process of disciplination, in that the reader is instructed not simply to look through the lens in different ways, but to do so according to a strictly prescribed time-based performance to which not only the eye, but also the hands and arms, and indeed the whole body, took part. Within these prearranged settings the reader was further instructed to expect to see certain effects at given moments, in particular the inversion of the image. What is most important, the visual effects linked to the two definitions of the "point of inversion" were described as manifesting themselves at different moments: first the rays converged at the "centre of inversion", and the lens became dark, and only later, when the lens was in the middle

[47] "Hanc operam supervacaneam non iudicabis, nec opus reiterare pigeat, non festinando, sed sensim et sedulo, ex hac disquisitione facilitas dependent" (Della Porta 1962, 103).

[48] "Oportet igitur manu bilentem capiamus, ac prius eam supra oculum opponamus, ut magnitudinem longe collocatam, ita ut est, recta, clara et paulo maior videatur. Mox specillum paulatim ab oculo recedat et recessum paulo ampliorem magnitudinem repraesentabit, et sensim eundo, ita turgidiour fiet, ut vix specillo complecti possit, post, tabescentie forma, pervertitur et demutatur ut specillum tenebris obducatur, appulsu e refractorum radiorum multitudine in ipso inversionis centro. Paulatim deinde specillum elongando manibus, ut inter visum et rem visam medium interiaceat, evanescunt tenebrae et magnitudo conversa, in angustum orbem collecta et quae ubiubi clarior videbitur. Et antrorsum iterum manus elongando crassescit moles, usque donec iterum totum specillum occuparit. Denuo pellitur imago, et transvolat, et tenebrae glomerantur et conglobantur in puncto inversionis: nam fieri non potest de recto ad inversum transmigratio, nisi per medium. Demum ulterius procedendo pervenit in vicinia magnitudinis, ubi clara, aequalia recta ut erat prope oculum conspicietur" (Della Porta 1962, 103).

between eye and object, the image appeared inverted. Once the reader had gone through this virtual experience, he or she was requested to go through the performance once again, this time conceptualizing it in geometrical form with the help of a diagram (Fig. 4.7):

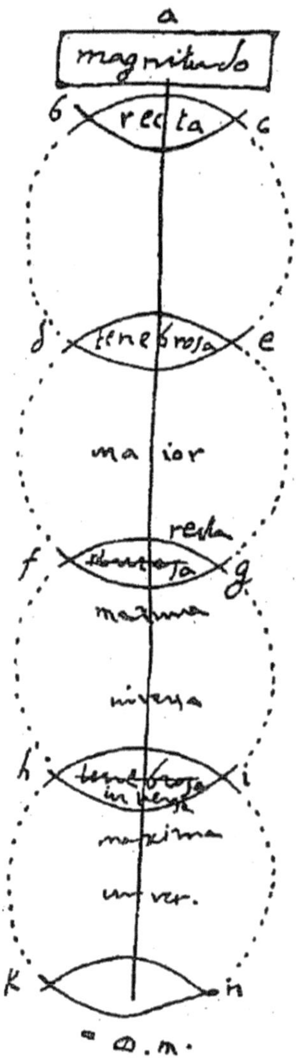

7 (DT 3A 2)

Fig. 4.7 *De telescopio*, chapter 3A, proposition 2

> Having seen those things, we have to go further and we shall here expound an example, so as to make what we have already explained in words even clearer by drawing lines. Let an object A be seen by an eye located in the position indicated by M. We label the lens, which is posed above the eye, with the annotations K N. We shall see the object indicated by A as it is in its place, upright, clear and of the same dimension. Now let us move away the lens, but not much and you will see the object become larger, and when you will se the object become so large that the lens cannot contain it any more, darkness will spread all over the lens. This happens exactly at the point of inversion and we shall label this place with the annotations H I. Proceeding forward the object expands, and when we arrive in the middle F G between object and eye, the image will become inverted, contract, and be seen somehow clearer. When the lens moves further away, the object appears vaguer, until the image will cover the lens with darkness, and we shall label this place with the annotations D E, as it corresponds to the upper point of inversion. When the lens once again moves away, the object becomes smaller and appears closer to the lens and will be seen clear, upright and as it is.[49]

The combination of verbal descriptions and diagrams endowed the experience with a geometrical structure in which the positions of the lens at which specific effects were observed became geometrical segments labelled by two "annotations" ("notulis"), creating a bridge between diagrammatic descriptions and empirical, operational ones. As I will discuss more in detail later on, Della Porta's strategy can in my opinion be seen as a key stage in a process of "mathematization" of experience. The second, hybrid description also communicated the scale at which the experience obtained, since the points of inversion were represented as having a mutual distance equal to twice the diameter of the lens' faces. The gradual transition between experiences and diagrams also helped understand that the latter were only a schematized formalism, and that no complete coherence could be expected between the senses and geometrical representation, as a number of ambiguities suggested: (1) the experience of image inversion did not obtain at the so-called point of inversion; (2) proposition DT 3A 1 had shown that no unique point of inversion existed, but in Fig. 4.7 a point of inversion linked to ray focussing was identified and located near the circumference of the lens' circles; (3) later on, the point of inversion was drawn in diagrams at the centre of the spherical surface. These discrepancies did not bother Della Porta: other than for Maurolico, Kepler or Galileo, for him geometry was not a rigid language of nature to be followed beyond experience to formulate proofs in Euclidian style, but rather a flexible tool for relating experience with itself, so that approximations were to be taken into account.

---

[49] "His perceptis, age ulterius procedamus, et illustremus exemplo, ut quae verba iam recensita sunt, lineamentorum ductu clariora evadant. Esto visenda magnitudo A e regione oculi quam M litera indicabit: specillum notulis obsignabimus K N, quod supra oculum constitutur. Hinc magnitudo, quam A notula obsignavimus, ut est suo loco recta, aequalis et nitida conspicietur. Mox specillum ulterius, sed non longe removes, videbis magnitudinem incrementum suscipere, et ubi maiorem concepisse magnitudinem comperies, ut specillum maiorem concipere non valeat, tota oculorum acies tenebris offundetur, quod eveniet in ipso inversionis puncto, locum H I notulis decorabimus. Dein ultrius procedendo, moles adhuc expanditur, et cum inter utrumque medium magnitudinis et oculi venerit FG convertitur imago, contrahitur et ubivis clarior cernitur. Abeat longius specillum, et molem laxiorem videbis, usque donec imago specillum tenebris obumbrabit, locum DE notulis configurabimus, quod in puncto inversionis superiori constituitur. Tandem ulterius abeundo angustior fiet magnitudo, et magis inhaerendo, et clara, recta, et ut est conspicabitur." (Della Porta 1962, 103).

4.9 Biconvex Lenses in Draft Chapter 3A of *De telescopio* (DT 3A)

After the reader had been introduced to the geometrical point of inversion and to visual image inversion, Della Porta proceeded to demonstrate how the two were related: using a new version of his rules for image construction in the lens, he built diagrams for each of the effects described in words, such as inversion and magnification. As anticipated, he employed a cathetus line passing through the centre of the upper circle, which now also factually corresponded to the point of inversion. In DT 3A the cathetus therefore retained the innovative feature that it had acquired in DT 3C, namely the fact that it could carry the image of the object. The propositions followed the order in which the various effects had appeared in the initial description: when eye or object are very near the lens, the image is equal to the object (DT 3A 3), when the eye is above the point of inversion, the object is upright and is larger if the eye is nearer to the lens (DT 3A 4 and 5); when the eye is below the point of inversion, the image is inverted and larger if the eye is further from the lens (DT 3A 6 and 7). Statements on other specific effects follow (e.g. lens obscured, image seen double, semilenses), and in particular one discussing an image "hanging in the air" (DT 3A 11). In the following, I will discuss only how Della Porta dealt with image inversion (DT 3A 6) and images hanging in the air (DT 3A 11).

Proposition DT 3A 6 (Fig. 4.8) described how the image of an object located above the point of inversion appears inverted. The text explained how the ray from A is refracted along the path ABCD, and the one from B along the path BFHD,

Fig. 4.8 *De telescopio*, chapter 3A, proposition 6

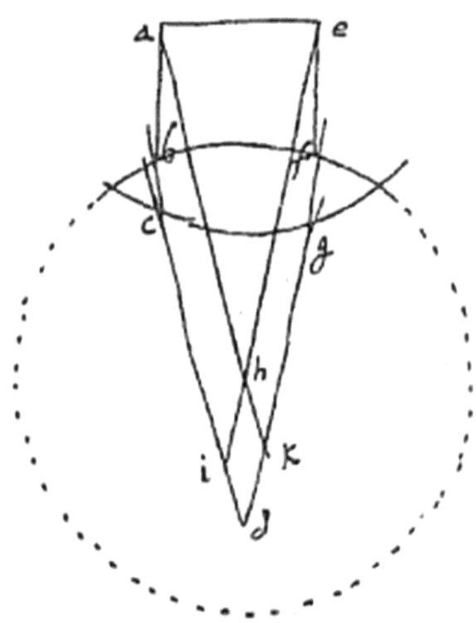

8 (DT 3A 6)

where D is the eye. The catheti are drawn through the centre E of the upper circle, and the cathetus from A intersects the ray from B and vice versa, forming the inverted image IK. As was the case in DT 3C, the refracted rays do not cross each other, but the image is nonetheless inverted. Della Porta explained:

> [The inversion occurs] because the cahteti going out from the centre always cut the lines carrying the images in a crossed way, so that the part from the right goes to the left, and the one from the left goes to the right, so that they bring an inverted image of the object.[50]

Here the similar functions of the cathetus and of the ray are explicitly spelled out, although the rays are still characterized as "lines carrying the forms [of the image]" ("lineae formas deferentes"). At this stage ray and cathetus have become equivalent tools to geometrize optical experience. While traditional optical diagrams were primarily drawn to represent some theory of how image is formed, and were usually drawn for those cases particularly interesting from the point of view of the theory, Della Porta's diagrams were rather constellations of lines and points which were expected to fit as many experiences as possible, while abiding the same rules. It was a process of bottom-up modelling whose outcome was a tacit but radical detachment from the world of perspectivist optics. It was thanks to this detachment that the notion of point of inversion could be adapted to lenses not just as a sterile, abstract notion resulting from a geometrical construction, but as a frutiful conceptual tool to grasp and further investigate visible phenomena.

Despite his "geometrization," Della Porta also distinguished between effects that, while corresponding to the same diagram, nonetheless appeared different to the eye. In proposition DT 3A 11 he discussed the formation by lenses of images "hanging in the air": "In biconvex lenses, when the eye is below the point of inversion, the image will be seen as inverted and hanging in the air".[51]

The relevant diagram (Fig. 4.9) was exactly like that of proposition 6, as far as ray path and catheti were concerned, but the two propositions were kept separate, as DT 3A 6 referred to an inverted image seen by an eye looking straight into the lens, while DT 3A 11 to an image "hanging in the air, inverted, outside of the lens" which however "the eye, if it moved upwards, could still see unchanged at the same location FG".[52] Here we encounter a very interesting ambiguity which already characterized images hanging in the air produced by the glass sphere: the construction of the image of DT 3A 11 required specifying the position of the eye, to which the refracted rays converged. However, once the image had been located, Della Porta

---

[50] "Quia catheti ex centro exeuntes semper lineas formas deferentes transversae decussant, ut dextera pars in sinistram vergat, et sinistra in dexteram, ut magnitudinis imaginem conversam referant" (Della Porta 1962, 106).

[51] "In convexis utrinque specillis, oculo sub inversionis puncto statuto magnitudo transversa et pendula in aere conspicabitur" (Della Porta 1962, 108).

[52] "in aere pendulam et trasversa extra specillum", "at si oculus altius subeet in FG suo loco transversa videbitur ut est" (Della Porta 1962, 109).

Fig. 4.9 *De telescopio*, chapter 3A, proposition 11

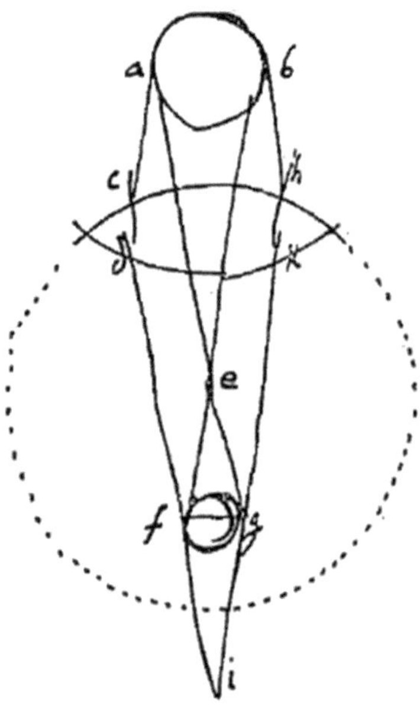

9 (DT 3A 11)

stated that it was possible to continue seeing it unchanged even when moving the eye, something that would have been impossible for the image in proposition DT 3A 6. Clearly, Della Porta had both experienced and correctly grasped the different properties of those two kinds of optical effects which are today conceptualized in terms of "real" and "virtual" image. He was trying to capture the difference in verbal terms, and, to do that, he had to modify the physical significance of the ray, which could form an image even without reaching the eye. While he had recognized that the two images in DT 3A 6 and 11 corresponded to the same ray-tracing constellation, he expressed their different quality by representing the object in the diagram of DT 3A 11 as a circle instead of as a line, perhaps suggesting three-dimensionality.[53] Thus, diagrams could also serve as hybrid media combining constructions drawn according to predetermined rules with more pictorial features which only conveyed a meaning in combination with the textual component of the proposition.

[53] Della Porta 1962, 123 (3B) and 135 (3C).

4.10 Summary and Conclusions

In the previous sections I have endeavoured to reconstruct Della Porta's strategy for investigating the formation of images in biconvex lenses as employed in the treatise *De refractione* (DR) and in the three drafts of a chapter on biconvex lenses from the unfinished treatise *De telescopio* (DT). I have argued that Della Porta's treatment of (biconvex) lenses followed and further expanded the methodology which the Neapolitan scholar had successfully deployed in studying reflection in spherical mirrors and refraction in the glass sphere. The key features of his approach were:

- a lack of committment to general views on light and vision, or at any rate no attempt to let such views influence the investigation of optical devices
- a focus on one specific optical device at a time (the spherical mirror, the glass sphere, biconvex lenses, etc.)
- a systematic attempt at describing in a standardized way all possible optical effects that could be produced by the device in question, individuating a small number of parameters relevant for all phenomena (e.g. distance between eye and device). The development of standardized descriptions was thus also a process of classification of the effects linked to a given device
- tentative formulation for each device of rules for reconstructing and/or predicting the visual effects it could produce

In all of these activities, hybrid communication strategies involving both texts and drawings were employed, and diagrammatic methods (i.e. diagrams and the rules for drawing them) had a key mediating and epistemic function. Moreover, notions and rules developed for simpler devices could serve as heuristic means for conceptualizing other, more complex ones ("thinking with objects"). In my discussion I have argued that diagrammatic methods took up a key epistemic function in Della Porta's work. In particular, the rules for the functioning of a given device were expressed in the form of rules for drawing the relevant diagrams and not as statements of some general physical principles. In this way diagrams became a medium for constructing new optical concepts (e.g. the point of inversion of lenses), and connect qualitative visual experiences to geometrical schemes. Diagrams also mediated between different devices, since the rules for one device might be adapted to another one. Finally, no overarching rules for all devices were formulated, or even assumed to exist, although individual concepts were tentatively carried over from one device to the other.

As an example of how these general features functioned in a practical instance I have discussed the way in which Della Porta adapted to lenses the notions of point of inversion and cathetus line. In today's optics the point of inversion is a well-defined construct, whereas the cathetus has no physical meaning. In the sixteenth century the opposite was true: the cathetus was a well-established optical notion, which was clearly defined in the case of reflection and of plane refraction. The burning point of a spherical mirror was also a traditional concept, althouhg its exact location and the problem of aberration only became widely known in the late sixteenth century. However, the "point of inversion" had only become important quite recently: Thanks to the availability of glass mirrors, image inversion had attracted attention and it had been noted that its occurrence could be related to the relative position of eye, object and lens, leading to the emergence of the qualitative notion of a "point of inversion". By the middle of the sixteenth century it had been remarked that the point of inversion and the burning point had approximately the same location, but the relation between the two remained rather unclear, as they were linked to quite different effects. In *De refractione* Della Porta had plausibly adapted these notions to refraction in the glass sphere, but was still striving to extend them to lenses.

In all his efforts Della Porta built upon the perspectivist theory of image formation, according to which the image of a point-like object is seen at the intersection of the reflected or refracted ray carrying the image and of the cathetus. Della Porta tried and largely suceeded in embedding the point of inversion in this traditional scheme, but to achieve this aim, in DR and in the three extant version of chapter 3 of DT (3A, 3B, 3C) he had to go through a number of redefinitions of the cathetus and of the point of inversion for biconvex lenses. I have summarized the steps of this trial-and-error process in Table 4.1 and illustrated them in Fig. 4.10a–d. In the table, I have listed separately different definitions and locations of the point of inversion because they do not always coincide, although they are very near to each other. Although the sources do not offer any explicit statement on the subject, I would like to suggest that for Della Porta and his contemporaries speaking of a "focal point" in no way implied that it would correspond to an empirically identifiable point-like location. For example, the fact that the spherical mirror and the glass sphere did not concentrate rays in a single point was known to Della Porta, but he still spoke of a focal point/point of inversion. In a similar way, the fact that two different phenomena (image inversion and combustion) could both be associated to the same small, yet not point-like area was enough to conceive them both as linked to the same ideal "point". It is not possible to discuss here this issue in any detail, but I would like to note that it represents a feature of "mathematization" and "quantification" as practised in the early modern period – a feature which sets those practices apart from the modern ideal of an exact numerical correspondence between measurement and computation.[54]

The various successive definition of the cathetus and of the point of inversion were no "blind groping", as Naldoni claimed, but were motivated by the wish to deliver predictions consistent with as many observations on the lens as possible. The initial cathetus definition of DR VIII and DT 3B was abandoned because it was seen to contradict experience. In DT 3C and 3A Della Porta continued his attempts to integrate the point of inversion in the rules for drawing diagrams in such a way, as to be able to use it as a parameter in formalizing the observed connections between image inversion and space location of object and eye. His solution was to define cathetus and point of inversion in such a way, that the former passed through the latter by construction, so that cathetus crossing automatically occurred as soon as

[54] Frangsmyr et al. 1990, Garber 1999, Wise 1997.

Table 4.1 Summary of Della Porta's changing rules of image construction in biconvex lenses

	DR VIII (Fig. 4.10a)	DT 3B (Fig. 4.10b)	DT 3C (Fig. 4.10c)	DT 3A (Fig. 4.10d)
"Point of inversion": geometrical definition	Focus of rays	–	Focus of rays	Focus of rays
"Point of inversion": geometrical location	–	–	A segment, not a point	A segment near the centre of upper or lower circle
"Point of inversion": visual definition	–	Object position at which image inversion occurs	–	Eye position at which the image fills the lense
"Point of inversion": visual location	–	Along upper semidiameter of lower circle	–	At a diameter's distance from the lense
"Point of inversion": location in diagrams	Between centre of curvature and vertex of the lens	Above lower circle or below upper one	Slightly above lower circle or below upper one	Centre of upper or lower circle
Centre of cathetus lines: verbal definition	Centre of lower circle	Centre of lower circle	*Not* the centre of the lower circle – point of inversion	Centre of upper circle (= lower point of inversion)
Centre of cathetus lines: position in diagrams	Centre of lower cirlce	Centre of lower circle	Lower point of inversion	Centre of upper cirlce (= lower point of inversion)
Diagrammatic rule for image inversion	Ray crossing	Ray crossing	Cathetus crossing	Cathetus crossing
Statement of the relationship between image inversion and point of inversion	–	DT 3B 4 (later labelled "wrong"): image inversion *never* occurs when the object is above the point of inversion	DT 3C 6 and 7: if the eye is below/above the point of inversion, the image is inverted/upright – unless the object is very near the lens (DT 3C 3)	DT 3A 4 and 6: if the eye is below/above the point of inversion, the image is inverted/upright – unless the object is very near the lens (DT 3A 3)
		DT 3B 9: image inversion *always* occurs when the object is above the point of inversion		

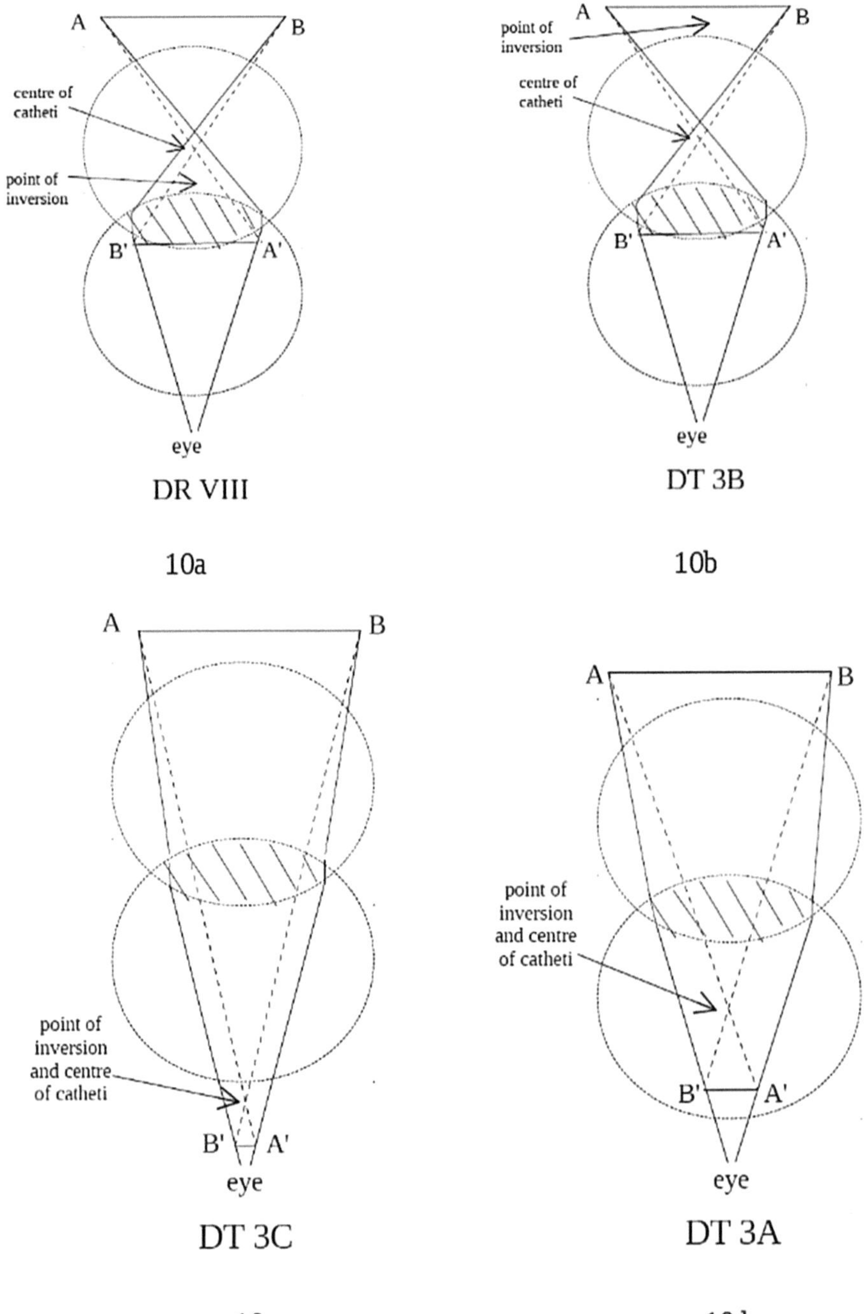

Fig. 4.10 **(a–d)**: Schematic summary of Della Porta's location of point of inversion and cathetus in *De refractione*, *De telescopio* Chapter 3B, 3C and 3A. AB is the object, A′B′ its image, the *full lines* are the rays and the *dotted lines* are the catheti

the eye was below the point of inversion. This diagrammatic rule allowed him to fulfil his goal of providing some general way of determining when image inversion occurred, provided that the cathetus (and not the refracted ray) was assumed to carry the image, so that cathetus crossing could give rise to image inversion. Della Porta's idea of inverting the image through cathetus crossing instead of ray crossing was extremely innovative, and perhaps it broke the rules of traditional optics even more than Kepler's overall rejection of the cathetus. As I have endeavoured to show, it was prompted exclusively by the need to find a geometrical model for lens refraction which fit experience, and Della Porta never offered a justification for it. Indeed, he did not even comment on the fact that his cathetus line had now acquired very singular physical features. However, he had managed to find constructions which semi-qualitatively fit both experience and the refraction rules for spherical surfaces which he had formulated in DR. In this process of knowledge production the cathetus, the ray and the point of inversion were employed as flexible "paper tools" to conceive and geometrize new optical experiences. Della Porta recognized the strength of his method, and in the beginning of DT 3A he offered a most impressive, rhetorically skilled presentation of how the experiences possible with a biconvex lens could be, and indeed should be, both perceived and conceived in terms of his rules for the biconvex lens.

The importance of developing new strategies to mediate new knowledge can hardly be overestimated, especially in the study of the early modern period, when so many of the new medial strategies of natural philosophy emerged: diagrams, mathematical notations, numerical tables, instrument-aided geometrical constructions, anatomical illustrations, standardized measuring units and devices, self-inscribing instruments, models – and new verbal strategies. Particularly challenging was the problem of how to make the procedures and outcomes of experimental research known beyond the immediate audience. In this respect, Della Porta displayed outstanding skills. They were in a sense the same skills he deployed in his theatre plays and in the most flameboyant natural magical experiments: a gift for grasping the attention of the audience and producing in them a sensual impression corresponding to the chosen experience. However, these were also the skills required to compose a pedagogically productive presentation. In the case of optics, a hybrid medial strategy combining words and diagrams was employed to mathematize visual experience.

The term "mathematization" is often understood to mean an attempt at formulating rigorous mathematical or geometrical laws of natural phenomena – or of nature in general, but I use it here in the sense in which Michael Lynch employed it when discussing how modern life scientists use photographs and diagrams to "represent" their observations and constitute their research object.[55] Lynch described how photographs of specimen were mutually compared and in turn represented in diagrams. He distinguished two steps in the visualisation process: "selection" and "mathematisation": "Selection concerns the way scientific methods of visualization simplify and schematize objects of study. Mathematisation concerns how such methods

[55] Lynch 1988.

attribute mathematical order to natural objects".[56] Lynch explains how scientists, starting from photographs of specimen tissues, choose specific aspects of them to be highlighted and diagrammatically represented with the help of lines, points, or grids. While the photograph is already an elaborated representation of "nature", Lynch argues that a diagram does not just simplify or schematize the photograph:

> [a diagram] brings into relief the essential, constant, veridical, and universally present aspect of the thing 'itself' [...] Both photograph and diagram exist on a common textual surface, and as such depend upon the artifices of inscription and interpretation while representing some wordly object.[57]

I would like to argue that, in our case, the verbal description of experiences with lenses at the beginning of 3A corresponds to the photograph, as a intersubjectively standardized, qualitative visualization of experience, while its diagrammatic representations play the same role as the diagrams Lynch discusses:

> [Diagrams are] visual documents [which] integrate the substantive, mathematical and literary resources of scientific investigation, and create the impression that the objects or relations they represent are *inherently* mathematical [...] The details of laboratory work, and of the visible products of such work, are largely organized around the practical task of constituting and "framing" a phenomenon so that it can be measured and mathematically described. The work of constituting a measurable phenomenon is not entirely separate from the work of measurement itself.[58]

Della Porta's mathematisation was not expected to deliver mathematical principles from which predictions could be rigorously deduced. Its aim was to employ mathematical-geometrical elements (e.g. numbers, points, lines, rotations...) to mediate experience and thus allow to conceive and manipulate it in new, hybrid ways. These manipulation methods might include both traditional mathematical and geometrical procedures, as well as other verbal, instrumental, pictorial or performative rules, as was the case with Berzelian fomulas of chemistry. The geometrical elements then became paper tools. While the physical-optical results of Della Porta's work on lenses may not amount to much in a presentist view, his modus operandi shares many traits with modern laboratory work, and I would like to claim that it constituted a major template for experimental philosophers of the following decades in their attempts at quantifying and mathematizing experience. It is also in this sense that Friedrich Steinle describes as "geometrization" and "mathematization" the way in which Andrè-Marie Ampère and Michael Faraday subsumed their experiences with electric and magnetic instruments:

> The initial phase of Ampère's research is a very different example of the process of mathematisation. His goal was not mathematising, but rather formulating the regularities of spacial processes. Since he had at his dispositions no concepts which could serve that goal, he had to develop himself mixed, geometrico-physical concepts, which were later generalised in an even more abstract direction in pure geometry [...] The invention by Faraday of the concept of line of force constitutes a further case of that type, in which the search for regu-

[56] Lynch 1988, 201.

[57] Lynch 1988, 211.

[58] Lynch 1988, 217–218.

larities led to the elaboration of a concept which later on became an abstract mathematical tool.[59]

In the context of Della Porta's exploratory manipulation of lenses and diagrams a new physico-geometrical concept of point of inversion of a lens emerged. It was not the modern notion, in which the point of inversion is identified with the focal point, while the visual experience of image inversion is defined in purely geometrical terms by the construction of the "real" image. Even when he achieved empirically validated results with his ray-tracing, Della Porta has been criticized by historians for allegedly not having recognized them as the valuable "scientific" outcomes which they are taken to be today. However, Della Porta's aim was not to reduce experience to the shadow of some perfect mathematical-geometrical principles, but rather to bring mathematics to the aid of the natural magician, who wished to better understand and replicate (visual) effects. Therefore, connecting the "visual" with the "geometrical" characterization of the point of inversion was a main priority for him. In this sense, his work should not be compared so much with that of authors trying to formulate laws of light and vision, but rather with later results on the "theory" of lenses and optical systems. By embedding the point of inversion in the diagrammatic treatment of lenses in practice, and not only in principle, Della Porta provided a conceptual basis for the study of lenses. By showing how diagrammatic methods could function as paper tools he delivered a most valuable methodology for constructing images in lenses, linking the workings of a lens to the position of its focal point. As far as we know from extant sources, Della Porta was the first author to systematically exploit the epistemic potential of optical diagrams which, although of course present since Antiquity, only in the early modern period became increasingly important as an independent epistemic means of knowledge construction. In his contribution to this volume (Chap. 7) Albrecht Heeffer convincingly argues that optical diagrams constituted an independent epistemic tool for scholars like Kepler, Mydorge or Descartes, who approached the problem of finding a universal law of refraction as framed by optical diagrams, which for them de facto represented the phenomenon of refraction they were trying to conceptualize in mathematical form.[60] More precisely, Heeffer shows how their reflections were guided by one specific diagram which had appeared already in the writings of Witelo and Della Porta as a means to conceptualize optical experience in geometrical terms. While Della Porta had used this and other diagrams to find qualitative and semi-quantitative rules of how specific optical instruments worked in practice, for more mathematically

---

[59] "La phase initiale de recherche d'Ampère [...] constitue un example tout différent de processus de mathématisation. Son but n'était pas de mathématiser, mais de formuler les régularités de processus spatiaux. Dans la mesure où il n'y avait pas de concept à disposition qui puisse servire à cette fin, il dut developper lui-même des concepts mixtes, géometrico-physiques, qui furent ensuite généralises de manière ancore plus abstraite en géométrie pure. [...] L'invention du concept de ligne de force per Faraday constitue un cas ultèrieur de type similaire dans lequel la recherche de régularités [...] mena à l'élaboration d'un concept qui devint par la suite un outil mathématique abstrait" (Steinle 2011, 83–84).

[60] Heeffer 2017.

minded authors that diagram became a starting point for transposing optical discussion into the media of Euclidian proportions and of algebraic and trigonometrical formulas. Like Berzelian formulas, the diagram thus achieved an ambiguous status between a representation of processes of refraction and a tool to experiment with those processes beyond the limits of workbench practices. From this experimental practice emerged the notion of "refraction" as a fundamental phenomenon following geometrical laws which are independent from the optical system in which the phenomenon manifests itself. Finding such universal laws was not Della Porta's aim, but the different epistemic goals which optical diagrams could serve testify to the broad span of cultures of optics and of mathematisation in the early modern period. Della Porta's research practices provide insights in a culture of experimentation and tentative conceptualization which was neither devoid of mathematical-geometrical components, nor informed by the Euclidian ideals prompting Maurolico, Kepler or Descartes, and instead represented a third way, probably also shaped by artisan knowledge. Della Porta's bottom-up approach to theorizing was in methodological perspective very similar to the experimental practices which Steinle described in Ampère and Faraday as "exploratory experimentation".[61]

Although it is not possible to specifically claim that the *De telescopio* had great impact on natural philosophy, since it was never circulated, it provides a good example of a methodology that Della Porta employed in most of his works, particularly the later ones. It was a methodology which he also made known through personal contacts: a "natural magical" approach to experimenting and representing experiences whose main characteristics lived on in the experimental practices of the following centuries.

Acknowledgements The author wishes to acknowledge the support of the DFG through the projects BO 4062/2-1 (TU Berlin, "Early particle physics") and KFOR 1927 (Institude for Advances Studies on Media Cultures of Computer Simulation (MECS), Leuphana Universität Lüneburg).

Bibliography

Bellé, Riccardo. 2017. Francesco Maurolico, Giambattista Della Porta and their theories on refraction. In *The optics of Giambattista Della Porta (ca. 1535–1615): A reassessment*, ed. Arianna Borrelli, Giora Hon, and Yaakov Zik, 169–200. Cham: Springer.

Bertoloni Meli, Domenico. 2006. *Thinking with objects. The transformation of mechanics in the seventeenth century*. Baltimore: The John Hopkins University Press.

Borrelli, Arianna. 2011. Giovan Battista Della Porta's Neapolitan magic and his humanistic meteorology. In *Variantology 5. On deep relations of arts, sciences, and technologies*, ed. Siegfried Zielinski and Eckhard Fürlüs, 103–130. Cologne: Walther König.

———. 2014. Thinking with optical objects: Glass spheres, lenses and refraction in Giovan Battista Della Porta's optical writings. *Journal of Early Modern Studies* 3: 39–61.

[61] Steinle 2016.

———. 2015. Genesis des Gottesteilchens: Narrativen der Massenerzeugung in der Teilchenphysik. In *Erzählung und Geltung. Wissenschaft zwischen Autorschaft und Autorität*, ed. Safia Azzouni, Stefan Böschen, and Carsten Reinhardt, 63–86. Weilerswist: Verlag Velbrück Wissenschaft.

Coopmans, Catelijne, et al., ed. 2014. *Representation in scientific practice revisited*. Cambridge, MA: The MIT Press.

Della Porta, Giambattista. 1560. *Magiae naturalis sive de miraculis rerum naturalium libri IIII*. Anverse: Plantin.

———. 1597. *Magiae naturalis sive de miraculis rerum naturalium libri XX*. Frankfurt a. M: Marne&Aubry.

———. 1593. *De refractione optices parte libri novem*. Naples: Carlino&Pace.

———. 1962 [1610–1615]. *De telescopio*, ed. Maria Amalia Naldoni. Florence: Olschki.

———. 1610-1615. *De telescopio* [manuscript]. Biblioteca dell'Accademia Nazionale dei Lincei e Corsiniana, Roma, Archivio Linceo 14. Online version:http://bibdig.museogalileo.it/Teca/Viewer?an=000000935209.

Dupré, Sven. 2000. Mathematical instruments and the 'Theory of the concave spherical mirror': Galileo's optics beyond art and science. *Nuncius* 15: 551–588.

———. 2003. The dioptrics of refractive dials in the sixteenth century. *Nuncius* 18: 1–29.

———. 2005. Ausonio's mirrors and Galileo's lenses: The telescope and sixteenth-century practical optical knowledge. *Galilaeana* 2: 145–180.

———. 2006. Visualisation in Renaissance optics: The function of geometrical diagrams and pictures in the transmission of practical knowledge. In *Transmitting knowledge. Words, images, and instruments in early modern Europe*, ed. Sachiko Kusukawa and Ian Maclean, 11–39. Oxford: Oxford University Press.

Eamon, William. 1994. *Science and the secrets of nature. Books of secrets in medieval and early modern culture*. Princeton: Princeton University Press.

Frangsmyr, Tore, J.L. Heilbron, and Robin E. Rider, ed. 1990. *The quantifying spirit in the eighteenth century*. Berkeley: University of California Press.

Galison, Peter. 1997. *Image and logic: A material culture of microphysics*. Chicago: University of Chicago Press.

Garber, Elizabeth. 1999. *The language of physics. The calculus and the development of theoretical physics in Europe, 1750–1914*. Boston: Birkhäuser.

Georgescu, Laura. 2014. The diagrammatic dimension of William Gilbert's De magnete. *Studies in History and Philosophy of Science A* 47: 18–25.

Gooding, David W. 1990. *Experiment and the making of meaning. Human agency in scientific observation and experiment*. Dordrecht: Springer.

Heeffer, Albrecht. 2017. Using invariances in geometrical diagrams: Della Porta, Kepler and Descartes on refraction. In *The optics of Giambattista Della Porta (ca. 1535–1615): A reassessment*, ed. Arianna Borrelli, Giora Hon, and Yaakov Zik, 145–168. Cham: Springer.

Ilardi, Vincent. 2007. *Renaissance vision from spectacles to telescopes*. Philadelphia: American Philosophical Society.

Kaiser, David. 2005. *Drawing theories apart. The dispersion of Feynman diagrams in postwar physics*. Chicago: University of Chicago Press.

Klein, Ursula. 2003. *Experiments, models, paper tools. Cultures of organic chemistry in the nineteenth century*. Stanford: Stanford University Press.

Lindberg, David. 1984. Optics in sixteenth-century Italy. In *Novità celesti e crisi del sapere*, ed. Paolo Galuzzi, 131–148. Florence: Giunti.

Lynch, Michael. 1988. The externalized retina: Selection and mathematization in the visual documentation of objects in the life sciences. *Human Studies* 11: 201–234.

Mumma, John, and Marco Panza, ed. 2011. Diagrams in mathematics: History and philosophy. *Synthese* 186.

Naldoni, Maria Amalia. 1946. L'ottica del Porta dal 'De refractione' al 'De telescopio'. *Atti della Fondazione Giorgio Ronchi* 1: 54–72.

———. 1962. Un manoscritto inedito di G.B. Della Porta. In *De telescopio*, Della Porta, Giovan Battista, and Maria Amalia Naldoni, 21–28. Florence: Olschki.

Piccari, Paolo. 2007. *Giovan Battista Della Porta. Il filosofo, il retore, lo scienziato*. Milan: Franco Angeli.

Pickering, Andrew. 1995. *The mangle of practice: Time, agency, and science*. Chicago: University of Chicago Press.

Ronchi, Vasco. 1962. Du "De refractione" au "De telescopio". In *De telescopio*, Della Porta, Giovan Battista, and Maria Amalia Naldoni, 1–19. Florence: Olschki.

Saito, Fumikazo. 2008. Experimento e matemática: o estudo das lentes segundo a perspectiva de Giambattista della Porta (1535–1615) *Circumscribere* 4: 83–101.

———. 2011. *O telescópio na magia natural de Giambattista della Porta*. São Paulo: EDUC/ Livraria da Fisica Editorial.

Smith, A. Mark. 2015. *From sight to light: The passage from ancient to modern optics*. Chicago/ London: University of Chicago Press.

Steinle, Friedrich. 2016. *Exploratory experiments. Ampère, Faraday and the origins of electrodynamics*. Pittsburgh: University of Pittsburgh Press.

———. 2011. La mathématisation: avec ou sans mesures? Le cas de l'électromagnétisme. In *La mathématisation comme probleme*, ed. Hugues Chabot and Sophie Roux, 59–88. Paris: Éditions des archives contemporaines.

Wise, M. Norton. 1997. *The values of precision*. Princeton: Princeton University Press.

Wright, Aaron Sidney. 2014. The advantages of bringing infinity to a finite place: Penrose diagrams as objects of intuition. *Historical Studies in the Natural Sciences* 44: 99–139.

Zik, Yaakov, and Giora Hon. 2012. Magnification: How to turn a spyglass into an astronomical telescope. *Archive for History of the Exact Sciences* 66: 439–446.

Zik, Yaakov and Giora Hon. 2017. Giambattista Della Porta: A magician or an optician?. In *The optics of Giambattista Della Porta (ca. 1535–1615): A reassessment*, ed. Arianna Borrelli, Giora Hon, and Yaakov Zik, 39–58. Cham: Springer.

Chapter 5
Giambattista Della Porta's Theory of Vision in the *De refractione* of 1593: Sources, Problems, Implications

A. Mark Smith

Abstract In both the *Magiae Naturalis Libri XX* of 1589 and the *De Refractione* of 1593, Giambattista Della Porta likens the eye to a camera obscura, the pupil serving as a narrow "window" through which images from outside are projected onto the screen of the crystalline lens. This, as Kepler recognized, was a major insight. Yet even a cursory look at Della Porta's account of vision in the De Refractione shows that, far from being guided by this insight, he ignored it completely, the result being that his theory of sight conflicts with the idea that the eye acts like a camera obscura. It is my aim in this paper to explain precisely how the two conflict and, in the process, to suggest some possible sources and implications of that conflict. In the process, I will look at various aspects of his theory of sight with an eye toward explaining how they are at times both innovative and problematic.

After presenting the theory of retinal imaging in chapter five of his *Ad Vitellionem Paralipomena* (Kepler 1604), Johannes Kepler looks at how previous thinkers attempted to explain the visual process. In the course of his examination, he singles out Giambattista Della Porta for having demonstrated once and for all that vision is due to incoming light alone, not to something emitted by the eye. "Indeed," he enthuses,"thou hast blessed us, O excellent initiate of nature" by resolving once and for all the issue of whether sight depends on the intromission or extramission of rays (Kepler 1604, 209; translation from Donahue 2000, 224).[1] No less important to Kepler, however, is that Della Porta understood the function of the pupil in the transmission of images into the eye. In acknowledgment, Kepler quotes the relevant passage from Della Porta's *Magiae Naturalis Libri XX* of 1589:

> It is therefore clear to natural philosophers and optical theorists where vision takes place, and the question of whether vision occurs by intromission, which was so vexed in antiquity, is put to rest, and it could not be demonstrated by any other device. The image is transmitted

[1] Equidem beasti nos, o excellens naturae mysta; lite illa dirempta, receptione an emissione fiat visio.

A. Mark Smith (✉)
Department of History, University of Missouri, 101 Read Hall, Columbia, MO 65201, USA
e-mail: SmithAM@missouri.edu

A. Borrelli et al. (eds.), *The Optics of Giambattista Della Porta (ca. 1535–1615): A Reassessment*, Archimedes 44, DOI 10.1007/978-3-319-50215-1_5

> through the pupil, which is like a hole in a window, and the portion of the crystalline sphere located in the middle of the eye takes the place of a tablet, which I know will greatly please those of superior intellect. This has been shown at greater length in our *Optics* (Della Porta 1589, 267).[2]

In short, Della Porta apparently understood that, like the aperture in a camera obscura, the pupil causes the incoming image to be projected in reversed and inverted order to the front surface of the lens, which acts as a screen for the image.

As promised at the end of the long quotation, Della Porta elaborates on this point in his "Optics"—that is, the *De Refractione* of 1593—which was obviously well in the works or even finished by the time he published the *Magiae Naturalis* 4 years earlier.[3] In the first proposition of book 4 of the *De Refractione*, for instance, he claims that, "just as the light shining through a narrow window onto paper held near it represents objects illuminated by the sun, so it depicts the images of visible things [by shining] through the opening of the pupil and falling onto the crystalline humor" (Della Porta 1593, 91).[4] At two other places in the *De Refractione*, moreover, Della Porta suggests that the lens is not just a screen upon which images are projected though the pupil; it also obeys the same optical laws as a chunk of crystal or glass. Accordingly, in the prologue to book 2, which deals with how light is refracted through, and brought to focus by, spherical lenses, Della Porta informs us that a glass sphere "serves as a model of the eye, and the true locations of vision in the eyes cannot be determined by us unless we touch upon a few things about that [sort of sphere]" (Della Porta 1593, 35).[5] Implicit in this claim is that the crystalline lens serves as a focusing device, a point made more or less explicit later in the sixth proposition of book 3. "If someone were to look for a while with open eyes at the sun," Della Porta observes there, "and if the pupil were dilated, the sun shining down would ignite a fire at the back of the eye or would generate a great deal of heat" (Della Porta 1593, 75–6).[6]

It would seem, therefore, that Della Porta anticipated Kepler's theory of retinal imaging not only by likening the eye with its pupil to a camera obscura but also by recognizing that, rather than visually sense impinging rays of luminous color at its

[2] Hinc Philosophis et opticis patet, quo nam [Kepler = *quoniam*] fiat visio loco, ac intromittendi dirimitur quaestio, sic antiquitus exagitata, nec alio utrumque [Kepler = *unquam*] artificio demonstrare poterit. Intromittitur idolum per pupillam, fenestrae foraminis instar, vicemque obtinet tabulae crystallinae sphaerae portio in medio oculi locata, quod scio ingeniosis maxime placiturum. In nostris opticis fusius declaratum est. (translation mine).

[3] Della Porta refers to his "Optics" several times in the *Magiae Naturalis*, and the context within which he makes these references leaves virtually no doubt that he is alluding to the *De Refractione*; see, e.g., Della Porta, 1589, 269, 271, 277, and 279.

[4] Dico quod sicut corpora a sole illustrata per arctum fenestrae foramen in subiectum papyrum lumen repraesentat, sic idem visarum rerum spectra per pupillae foramen subiens in crystallinum depingit.

[5] Haec enim pila oculi specimen refert, nec vera a nobis loca visionum in oculis decerni possunt, nisi de ea. paucula quaedam praestrinxerimus …

[6] Tandem si quis apertis oculis aliquantisper solem intueretur, et pupilla panderetur, ex solis illapsu ignem in oculi fine accenderet vel ibi maximum ardorem ingereret.

front surface, the crystalline lens physically refracts them to the back of the eye. As we will see in the course of this paper, however, these apparent Keplerian insights into the workings of the eye had virtually no influence on the theory of vision Della Porta articulated in the *De Refractione*. In fact, his theory differed from Kepler's in nearly every significant respect. But not in *every* respect. We will also see that Kepler and Della Porta had one important thing in common: a profoundly critical attitude toward previous optical authorities in general and Alhacen and his three Perspectivist followers, Witelo, Roger Bacon, and John Pecham, in particular. Nontheless, despite deprecating the Perspectivists several times in the *De Refractione*, and despite manifesting a strong independent streak, Della Porta was heavily dependent on his medieval antecedents for the account of direct, unrefracted vision he provided in the third of the nine books comprising that treatise.

5.1 Della Porta on Unrefracted Vision

Let us start by comparing Della Porta's model of the eye, in the diagram to the left of Fig. 5.1, with that of Alhacen and the Perspectivists, in the diagram to the right.[7]

Despite obvious differences in the size and distribution of the various components, the similarities between the two models are evident. Like the Perspectivists, for instance, Della Porta assumed that the eye is spherical and composed of nesting

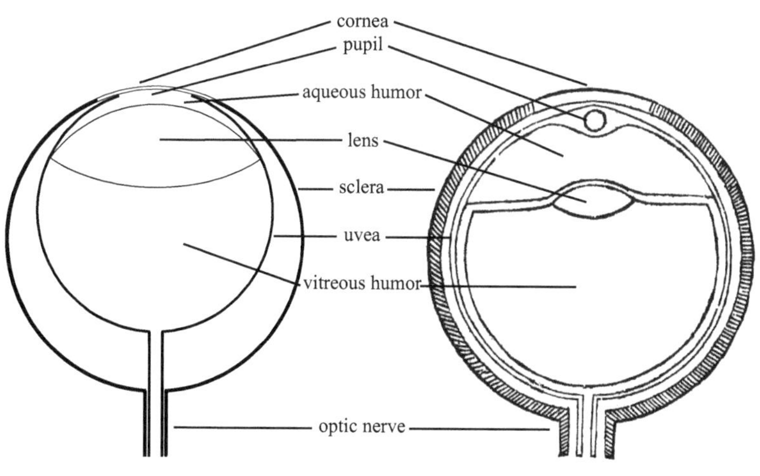

Fig. 5.1 Comparison of the Perspectivist model and Della Porta's model of the eye. For technical reasons, all images have been redrawn by the publisher. Their structure fully corresponds to that of the originals in Della Porta (1593)

[7] For Della Porta's diagram see Della Porta 1593, 68. For Alhacen's account of ocular anatomy, which was adopted by his Perspectivist followers, see Alhacen 2001, 11–22 (Latin), 348–355 (English).

tunics, although unlike them, he had the uveal tunic perfectly concentric with the sclera. Like the Perspectivists, as well, Della Porta located the crystalline lens toward the front of the eye, not at the center, where his anatomical source, Andreas Vesalius, placed it according to medical tradition.[8] And like the Perspectivists, Della Porta filled the three chambers of the eye with specific humors or fluids that differ in refractivity one from the other.

Della Porta also followed the Perspectivists in supposing that, after passing directly through the cornea along the orthogonal, individual rays from every point on the surfaces of external objects continue straight through the pupil and the front of the lens to converge toward the center of the eye. Taken as a whole, the resulting bundle of rays assumes the shape of a cone with its base defining the field of view and its vertex at the very center of the ocular sphere. Forming the cardinal reference point for spatial perception according to Perspectivist theory, this point constitutes the so-called center of sight.

As for Alhacen and the Perspectivists, so for Della Porta, this cone of radiation is crucial in determining not only what is seen directly in the visual field, but also how it is seen. Thus, in proposition 15 of the third book of *De Refractione* (Della Porta 1593, 85–6), Della Porta argues that, if BCDA in Fig. 5.2 represents the surface of a square object, its image will be projected into the eye within a pyramid whose edges all pass through the cornea at HG, the lens at FE, and the front of the lens LI toward center point M of the eye. There they would all converge, just as they would

Fig. 5.2 Della Porta's illustration of the cone of radiation

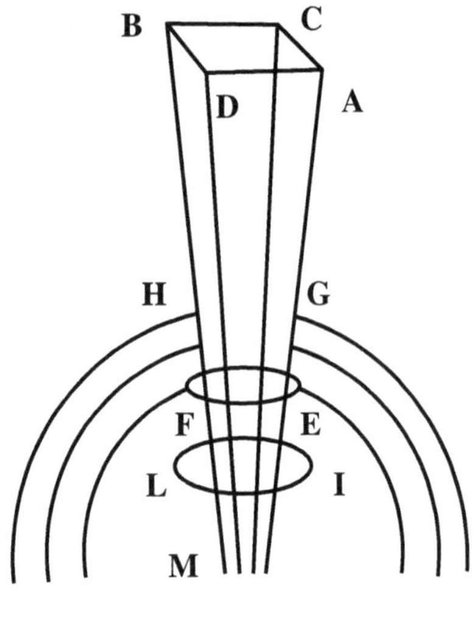

[8]Della Porta's description of ocular anatomy in book 3, prop. 1 is, for the most part, based on the one provided by Vesalius in book 7, chapter 14 of the *De Humani Corporis Fabrica*, although Vesalius locates the lens in the eye's center rather than toward the front, where Della Porta places it; cf. Vesalius 1568, 495–99, and Della Porta 1593, 65–68. Later, in book 3, prop. 14, Della Porta refers to Vesalius ("Vessalius") by name (Della Porta 1593, 84).

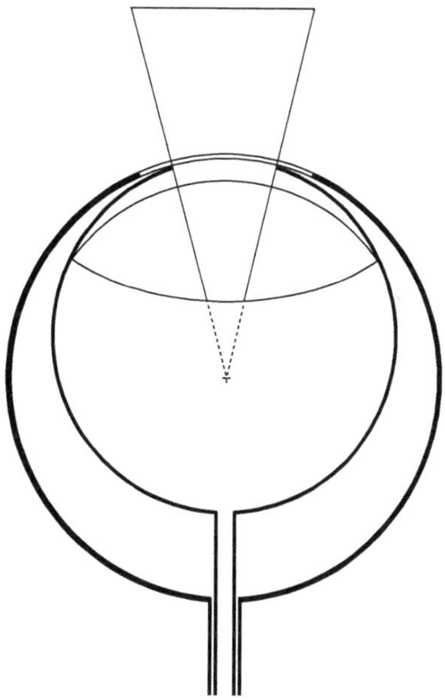

Fig. 5.3 The Perspectivist model of the cone of radiation

according to the Perspectivist model illustrated by Fig. 5.3. That they in fact do not coverge there is due to the refraction that occurs at the posterior surface of the lens and that channels all the rays within the cone into the hollow optic nerve at the rear. The image of the object is therefore conveyed to the eye, and eventually to the brain, as the object itself is actually oriented and located in the field of view. "This," Della Porta concludes, "seems to be vision in its truest sense, and so it appears from anatomy" (Della Porta 1593, 85).[9]

Based as it is on a cone of radiation, this visual model is clearly at odds with the model of retinal imaging articulated by Kepler in *Ad Vitellionem Paralipomena*. For one thing, it assumes that what amounts to a holistic image is projected into the eye according to a select group of particular rays emanating from single points on the object. Kepler's model, on the other hand, assumes that *all* the rays emanating from a given point on the object and reaching the lens, not just the perpendicular one, are implicated in the projection of the object's image into and through the eye. This point will become clear shortly. For another thing, despite his likening the pupil to the aperture in a camera obscura, Della Porta's model of vision, with its emphasis on the radiative cone, ensures that the image reaching the front of the lens and continuing through the eye retains its original orientation rather than suffering inversion.

Della Porta's conception of visual images also links him more closely to the Perspectivists, Roger Bacon in particular, than to Kepler. This linkage is especially evident at the terminological level, where Della Porta refers to such images variously

[9] Haec videtur germana visio, et ita ex anatome apparet.

as *simulacra, species, formae, idola, spectra, fantasmata*, and *imagines*. The sources from which Della Porta could have drawn these terms individually are legion, ranging from Pliny's *Natural Histories* and Seneca's *Naturales Quaestiones* to such contemporary works as Girolamo Cardano's *De Subtilitate* of 1550. But one possible source stands out: Roger Bacon's *De multiplicatione specierum*. A key point in that work comes at the very beginning, where Bacon informs us that the active agent in sight can take several names, including "'the similitude' [*similitudo*] of the agent, "'image' [*imago*], 'species' [*species*] 'idol' [*idolum*], 'simulacrum' [*simulacrum*] 'phantasm' [*fantasma*], 'form' [*forma*], 'intention,' 'passion,' 'impression,' and 'shadow of the philosophers,'" (Bacon 1983, 2–3). This conflation includes the full range of terms that Della Porta uses to designate "image" at various points throughout the course of his account of sight. Whether Bacon was the actual inspiration for Della Porta's own conflation of terms is an open question. We do know, however, that Della Porta was familiar with Bacon's *Perspectiva* (Bacon 1996), and the parallel between Bacon's and Della Porta's terminological conflations suggests that he knew the *De multiplicatione specierum*, as well. In both cases, moreover, he would have had access to the text only through manuscript sources, since neither work was committed to print before 1614 (Lindberg 1975, 38, 42). Whatever the case, there is no question that Della Porta followed Bacon, and the rest of the Perspectivists, in dealing with these variously-named images holistically rather than pointillistically.

5.2 The Function of the Pupil

To this point we have seen that Della Porta's basic theory of vision has far more in common with its Perspectivist antecedents than with the theory of retinal imaging that Kepler proposed about a decade after the *De Refractione* appeared. When we turn to certain analytic details, however, we find that Della Porta forged off in directions that put him in fairly sharp conflict with both the Perspectivists and Kepler. Just how far he diverged from both, particularly Kepler, is illustrated by his understanding of the pupil's function. In order therefore to set the appropriate background for an analysis of this difference, let us first look at how Kepler viewed the pupil and the part it plays in vision.

In Kepler's account of retinal imaging, the pupil has a dual role. On the one hand, it acts as a diaphragm. In that capacity, it narrows the sheaf of rays entering the eye from each point in the field of view, and by so doing, it enhances the brightness and clarity of the image projected on the retina by culling out extraneous radiation. On the other hand, in acting like the aperture of a camera obscura, the pupil causes all the incoming rays to reach the lens in inverted order. Consequently, on being refracted through the lens, all the rays within a given sheaf are brought to focus on the retina in the same inverted order. These points are easily understood according to the eye's basic structure and the way light refracts through lenses.

In the side-by-side diagrams of Fig. 5.4, for instance, AB represents the eye's lens, O1 and O2 being two luminous points radiating light to every exposed spot on

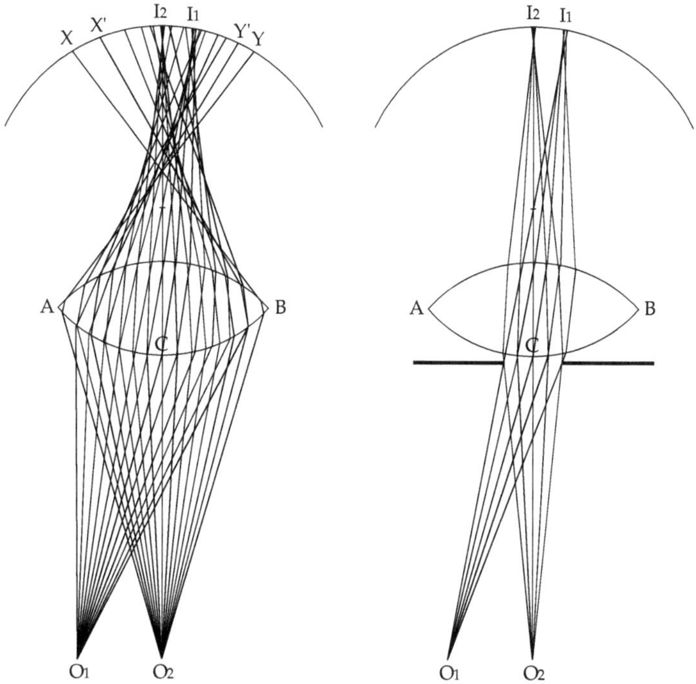

Fig. 5.4 Kepler's understanding of the function of the pupil

its facing surface. These two points thus form the vertices of radiative cones based on the front surface of the lens. On refracting through the lens, the rays from both points are subject to spherical aberration, which causes them to refract ever more sharply the closer to the edge of the lens they enter and exit it. Thus, in the left-hand diagram of Fig. 5.4, after passing through the lens, the ray from O2 striking closest to edge B will be so sharply refracted as to reach point X on the retinal arc at the back of the eye. Somewhat less sharply refracted, the next ray in line will reach point X', and so on. Likewise, the ray from O1 striking the lens closest to edge A will reach point Y on the retina, the next in line will reach point Y', and so on. Meantime, the rays that strike the lens close to point C at the middle of its anterior surface will refract less severely and will therefore aggregate at or near a focal point on the retina to form sharp images I1 and I2.

Because of spherical aberration, however, the clarity and brightness of those images will be compromised by the extraneous radiation that overlaps with them within arc XY. Thus, the extraneous radiation from point O1 will interfere with image I2, while that from point O2 will interfere with image I1, and this interference will be exacerbated by the extraneous radiation accruing from the formation of images at every other spot on the retina. As a result, the composite, point-by-point image of the field of view formed on the retina will be occluded by all the superfluous light projected onto the retina. The pupil's main function, according to Kepler, is thus to minimize such interference by sifting out all but the most central radiation because, after refracting through the lens, this particular radiation aggregates at or near the focal point to form the clearest image. Consequently, as illustrated by the right-hand diagram of Fig. 5.4, images I1 and I2 are projected upon the retina with optimal clarity because they are virtually free from the extraneous radiation that would occur if the pupil did not narrow the cones of radiation to relatively thin sheafs. Moreover, as is clear from the same diagram, those two images are cast on the retina in reverse order from that of their generating objects.[10]

Like Kepler (and to some extent the Perspectivists as well) Della Porta viewed the pupil as a light-filter that keeps the chamber enclosed by the uveal tunic adequately dark to render the images implicated in vision as bright and clear as possible for the visual faculty. Unlike Kepler, however, Della Porta ignored the pupil's role in sifting the radiation that strikes the eye, thereby reducing it to a thin sheaf before it reaches the anterior surface of the lens. What he did not ignore, though—and this sets him apart from both Kepler and the Perspectivists—is the natural tendency of the pupil to dilate and contract according light-intensity. Spurred on by the recognition of this phenomenon, Della Porta proposed two simple experiments to prove that the pupil's dilation and contraction is not just related to the intensity of light but contingent on it.

In proposition 6 of the third book, for instance, Della Porta suggests the following experiment based on the apparatus illustrated in the two diagrams of Fig. 5.5.[11] Take a ruler EFD about a foot long and a digit wide (i.e., a bit less than a centimeter wide), and cut a notch at E to accommodate the eye when that end of the ruler is brought right up to it. Shape two cylindrical pegs C and G, C being the smaller of the two. Then, holding the ruler right up to the eye and toward bright light, adjust the pegs along line ECG until the smaller one at C completely occludes the one at G. Fixing the pegs in that relative position, draw lines FA and DB tangent to their edges so as to mark out arc AB on the notch at E. Then do the same thing with the apparatus in dim light. If you compare arc AB formed in bright light (in the bottom diagram of Fig. 5.5) to its counterpart formed in dim light (in the top diagram), you

Fig. 5.5 Della Porta's instrument for judging pupil's reaction to light-intensity

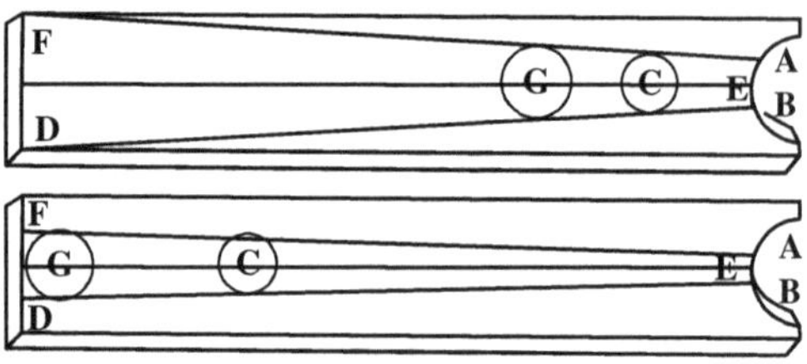

[10] For Kepler's complete explanation of the pupil's function as a diaphragm, see Kepler 1604, 183–215, esp. 210–214. For a somewhat more detailed explanation of Kepler's understanding of the retinal focus, particularly toward the sides of the retina, see Smith 2015, 354–63.

[11] For the full text of proposition 6, see Della Porta 1593, 74–6. Della Porta informs us that he based the apparatus illustrated in Fig. 5.5 on the one "that Archimedes used in determining the size of the sun" (quod Archimedes in dignoscenda solis quantitate usus est) in the early part of the *Sand Reckoner* (Della Porta 1593, 74).

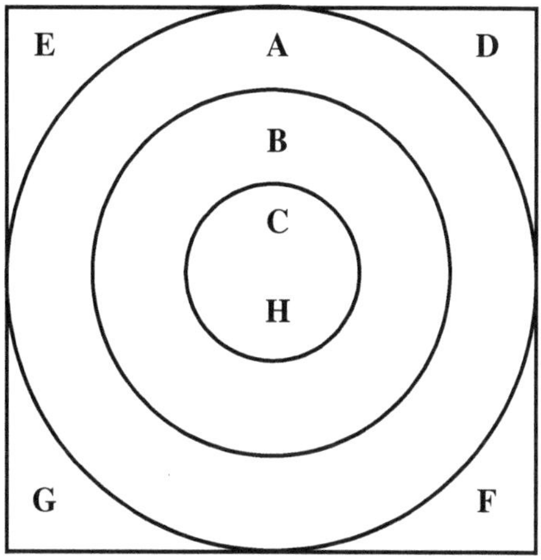

Fig. 5.6 Della Porta's instrument for judging pupil's reaction to vividness of color

will see that it is narrower, which indicates that the pupil was commensurately narrower in bright light than in dim light.

Another experiment to demonstrate the narrowing of the pupil in bright light, this time according to the vividness of color, is illustrated in Fig. 5.6, which accompanies proposition 8 of book 3. Let EDFG represent a plank one cubit (c. 1.5 ft = c.5 m) square. Drive a nail into its center at H so that the nail's head lies half a foot (c. 15 cm) above the plank. Start by painting the surface of the plank pitch black. Then position one eye right at the nail's head, and, looking straight down, place pebbles all around the edge of your field of view. Pulling your eye back, you will see that the pebbles form the large circle A in Fig. 5.6. Repaint the plank bright green, and repeat the procedure. This time the pebbles will form smaller circle B. Finally, paint the plank bright white, and repeat the procedure. This time the pebbles will describe the smallest circle C. All told, then, this experiment proves three things: first, that the pupil is consistently round; second, that it responds not just to the intensity of light but also that of color; and third, that it narrows as the colors become increasingly intense.[12]

Having shown experimentally that pupil size is inversely related to light intensity, Della Porta turns to the question of why this should be the case. He starts by noting that certain animals, such as owls, cats, and wolves, have naturally wide pupils. Other animals, such as hens, doves, and sparrows have naturally narrow pupils. What differentiates the two types of animals is their ability or inability to see at night. Some, in short, are nocturnal, others diurnal. It therefore follows that "those that are of a wide pupil [see] better at night and more weakly during the day, whereas those with a narrow pupil [see] more acutely during the day but scarcely anything at night" (Della Porta, 79).[13]

---

[12] For the full text of proposition 8, see Della Porta 1593, 77–9.

[13] Sed qui lata sunt pupilla noctu melius, interdiu hebetius, qui vero arcta interdiu acutius noctu vero nil penitus vident.

The reason is simple. On the one hand, the naturally constricted pupil of diurnal animals, including humans, allows them to function in broad daylight by reducing the amount of light that enters the eye. In so doing, it prevents the eye from being offended or injured by excess light while, at the same time, ensuring that the images entering the eye are not occluded by too much ambient light.[14] On the other hand, the naturally dilated pupil of diurnal animals allows as much light as possible to enter the eye during the hours of darkness, thus enhancing the brightness and clarity of the incoming images. Accordingly, nocturnal animals shy away from daylight because its brightness offends their eyes, thereby weakening their sight, whereas by damping the incoming light, the narrow pupils of diurnal animals, even at their most dilated, render those animals virtually blind at night (Della Porta 1593, 79–80).

Because Della Porta's account of night vision and the pupil's function in promoting or preventing it is so straightforward and sensible, and also because it conforms so closely to our modern understanding of the phenomenon, it is easy to lose sight of how innovative it was for its day. In particular, Della Porta flew in the face of virtually every classical and contemporary authority, who imputed the ability of nocturnal animals to see at night to the intensity of light shining from their eyes. Strong support for this supposition lay in the well-known fact that, unlike the eyes of diurnal animals, those of nocturnal animals glow quite brightly in the dark.

Fully aware of this phenomenon, Della Porta nonetheless denied that it was due to light shining from the eye because he was adamant that nothing whatever is emitted from the eye in the visual process. Consequently, in book 7, prop. 2 of the *De Refractione* (Della Porta 1593, 159–161), he offered a physical rather than a physiological explanation. Taking his cue from Aristotle (*De sensu* 2, 437a33-b10), he attributed the eye's bright glowing to the extreme polish of the crystalline lens, which shines by capturing and reflecting what little light enters the wide-open pupil. The apparent glittering is due to swift eye movement, which causes the reflected light to be dispersed in various directions so that the eye as a whole appears to scintillate. According to Della Porta, then, the phenomenon has nothing whatever to do with visual spirits or ocular luminosity; it involves the physics of light alone.

5.3 The Problem of Peripheral Vision

Della Porta's emphasis on physical over physiological and, even more to the point, psychological causation is especially evident in his account of peripheral vision, which departs sharply from its Perspectivist counterpart. The problem of peripheral vision is straightforward enough. As we saw earlier in Sect. 5.1, Alhacen and his Perspectivist followers explained vision on the basis of a cone of radiation defined

[14] The severity with which excess light can offend the eye, according to Della Porta, ranges from the creation of afterimages to snow-blindness or, at worst, the generation of intense heat or even fire at the back of the eye by the focusing of light through the crystalline lens, as mentioned at the beginning of this article.

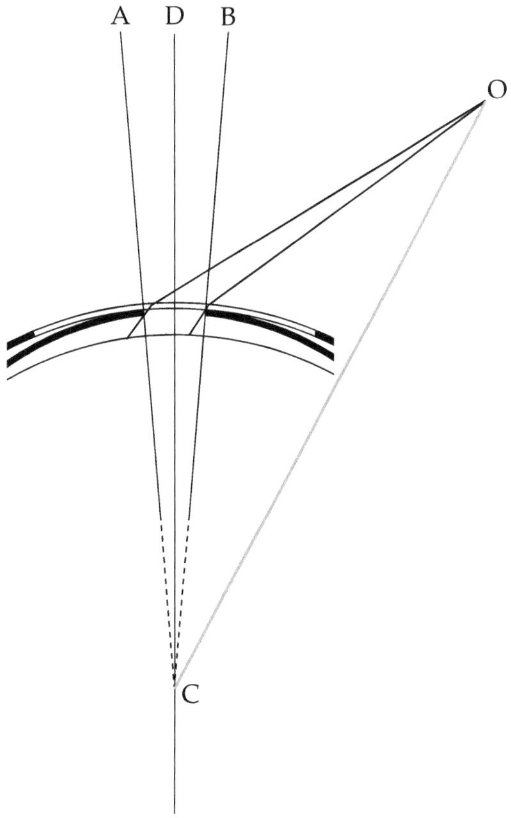

Fig. 5.7 Alhacen's model of peripheral vision

by the pupil and consisting of perpendicular rays converging toward the center of sight. Being perfectly direct, these rays make a strong enough impression on the lens's anterior surface to be visually sensed. All other impinging radiation, being oblique and thus relatively glancing, is too weak to make a sensible impression. The problem is that, being quite narrow, the opening of the pupil constricts the cone of radiation so severely that its base is not nearly capacious enough to allow for the scope of the actual field of view. Extending almost 180° along the horizontal, in fact, this field is well beyond the sweep of the visual angle dictated by the pupil, even at its most dilated.

Faced with this fact, Alhacen and his Latin followers were forced to modify their claim that only perpendicular rays are implicated in vision by bringing oblique radiation into the account. Thus, in Fig. 5.7, which represents a cross-section of the eye as understood by Alhacen, ACB is the cone of radiation with its vertex at center of sight C and its axis along CD. CD also serves as the so-called visual axis, which defines the line along which vision is clearest and most definite. Narrowed by the lens, this cone encompasses the relatively narrow field of view defined by angle ACB. Yet, although it lies well outside that field, point O can be seen. In order to explain this apparent anomaly, Alhacen assumes that all the light from O channeled through the pupil to the lens along oblique rays is sensed by the lens, albeit more weakly than the orthogonal rays striking it from points between A and B within the actual cone of radiation. In short, all the radiation within the oblique cone of radiation whose vertex is at O has a sensible effect on the lens's surface.

Furthermore, whereas the visual impressions made on the lens by orthogonal rays within cone ACB bear a one-to-one correspondence with the points from which those rays originate, the same does not hold for the cone of oblique radiation. Not just one but all the rays within that cone are felt by the lens, and the more oblique the impinging radiation, the weaker the visual impression, so the resulting composite visual impression is chaotic. How, then, do we end up getting a coherent view of the object? The answer, according to Alhacen, is that the visual faculty interprets the composite impression at the base of the oblique cone as if it were a single point-impression created along imaginary ray OC directed from the object to the center of sight C and therefore perpendicular to the anterior surface of the lens. Unlike the perpendicular rays within the actual cone of radiation, though, this one is virtual, not real. As such, therefore, it is a purely psychological construct. Since, moreover, this virtual ray does not correlate spatially to the actual radiation creating the composite impression of O on the lens, there is no point-to-point correspondence between that composite impression and the perceptible impression of O along virtual ray OC.

By positing such virtual radiation, Alhacen was able to extend the cone of radiation to the very edges of the visual field and, on that basis, to ensure that everything in the field is seen according to its proper side-to-side and up-and-down orientation. In short, Alhacen was able to avoid the possibility of image inversion due to radiation arriving at an extreme slant from points far to the side of cone ACB. He was also able to account for variations in visual acuity within the field of view according to the relative weakness or strength of the impinging radiation. Things far to the edge of that field appear nebulous because the rays emanating from them reach the lens at a sharp slant and make a commensurately weak overall impression. As they approach the center of the field of view, however, things appear increasingly clear because their radiation strikes the lens ever more directly. The clearest vision of all occurs within the compass of cone ABC according to two determinants. First, since each point within the field of view defined by cone ACB propagates its light to the lens along a perpendicular ray, that ray makes the strongest possible visual impression on the lens. Second, because the light from each such point also strikes the lens along oblique rays surrounding the perpendicular one, the point is seen along a virtual perpendicular ray that coincides with the actual perpendicular ray. Consequently, the impression along the real perpendicular ray is reinforced by that along its virtual counterpart.[15]

Taking an entirely different tack from Alhacen in his account of peripheral vision, Della Porta looked to the physics of refraction rather than the psychology of perception. In the first proposition of book 4, for example, he argues that, when we see an

[15] For Alhacen's full account of peripheral vision, see Alhacen 2010, 100–14 [Latin], 296–305 [English]; on the reinforcement of visual clarity within the cone of radiation, see ibid.,111–14 [Latin], 305–07 [English].

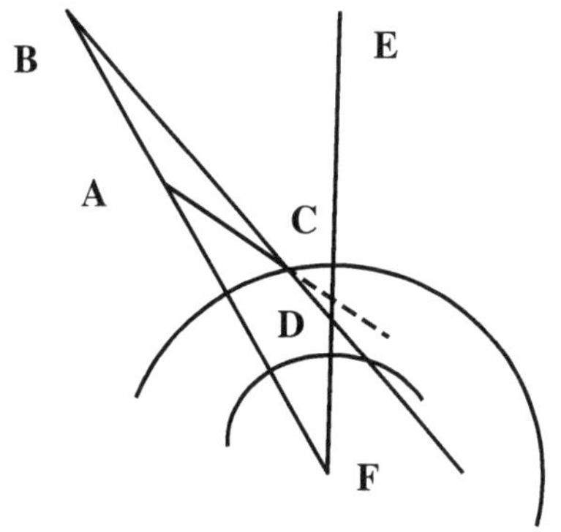

Fig. 5.8 Example of peripheral vision according to Della Porta

object off to the side, we do not see it where it actually lies because its radiation is refracted by the aqueous humor. This point is illustrated in Fig. 5.8, where the larger and smaller arcs represent the eye's outer surface and lens, respectively, both sharing center point F. Let A be an object lying toward the edge of the visual field, and let ray AC from it strike the cornea at point C, where it refracts along CD to reach point D on the lens through the pupil.

According to Alhacen's account, point A will be perceived along virtual ray FA extending from center of sight F, so it will be perceived to lie where it actually does with respect to the eye because FA is the perpendicular line of sight along which it is seen. Not so according to Della Porta. He claims that point A will be perceived at B, where the extension of refracted ray CD intersects the extension of line FA. In other words, it is not point A itself, but rather its image B that is seen, and it is seen according to the cathetus rule. This rule, which Della Porta establishes in book I, prop. 7, states that the image of any object seen by means of refraction appears at the intersection of the ray refracted to a given viewpoint and the cathetus of incidence dropped orthogonally from the object to the interface of refraction. It therefore follows that FAB in Fig. 5.8 is the extended cathetus of incidence, which is perpendicular to the outer surface of the eye, DCB is the extended refracted ray, and D is the viewpoint.[16] DCB, not FAB, is therefore the line of sight along which the image is seen. It also follows that, even though D is to the right of axis EF and thus on the opposite side of object A, image B is on the same side of visual axis EF as A, so A will be seen with its proper orientation rather than inverted.

Most noteworthy about this explanation, of course, is that it does away with the center of sight at the vertex of the radiative cone, this point serving as the cardinal viewpoint in the account of sight articulated by Alhacen and his Perspectivist followers. For Della Porta, any point on the lens that is open to impinging light can serve as a viewpoint, which is to say that for any eye there are as many viewpoints as there are points on the lens's anterior surface. This in turn implies that each such

[16] For Della Porta's argument in book 4, prop. 1, see Della Porta 1593, 91–2; and for his establishment of the cathetur rule in book I, prop. 7, see Della Porta 1593, 15–16. Note that, despite its placement in Fig. 5.8, D is meant to designate the point where the refracted ray intersects the anterior surface of the lens.

viewpoint on the lens must have a specific, corresponding ray associated with it. Take the example illustrated in Fig. 5.9, which is based on Fig. 5.8. As we saw in the previous paragraph.

Della Porta argued that object A will be seen at image location B, where the extension of refracted ray CD reaching viewpoint D intersects the extension of cathetus FA. But if every point on the lens can serve as a viewpoint, then when ray AC' is refracted to point D' on the lens, A ought to be seen from that viewpoint at image location B'. Since A is perceived at B and nowhere else, it follows that A can only be seen through incident and refracted ray couple ACD.

Why this particular ray couple and no other? Della Porta never answers this question explicitly, but proposition 9 of the fourth book suggests a simple mechanism for determining such ray-couples (Della Porta 1593, 103). In that theorem Della Porta invites us to think of ABCDE in Fig. 5.10 as an object viewed through pupil GF of the eye, whose lens is HLMI and whose center point is K. BCD will be seen according to a cone of radiation within which rays from D, B, and all points between them, including C, strike the lens perpendicularly and converge toward

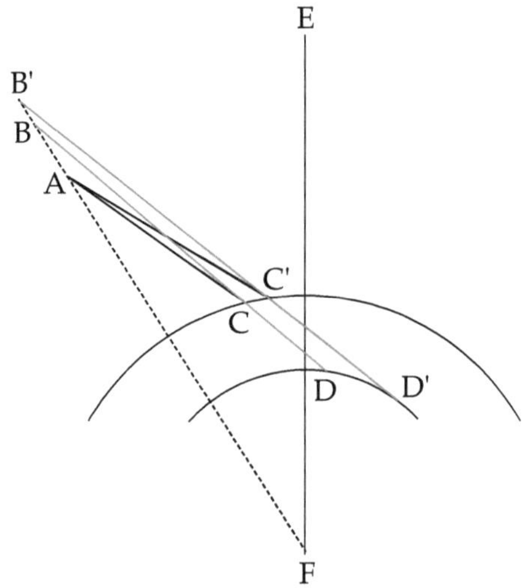

Fig. 5.9 Example of peripheral vision according to Della Porta

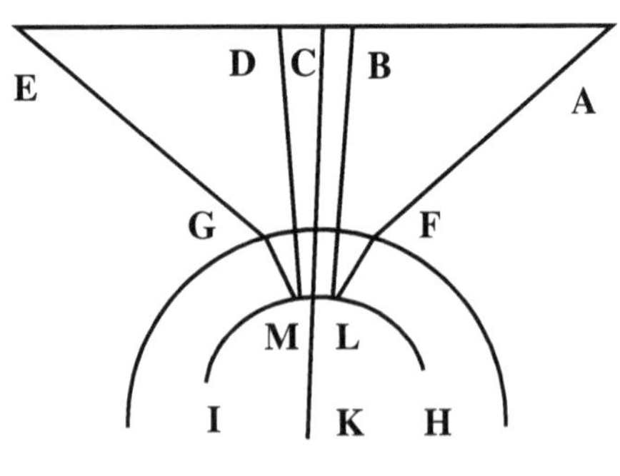

Fig. 5.10 Della Porta's "funnel" of radiation

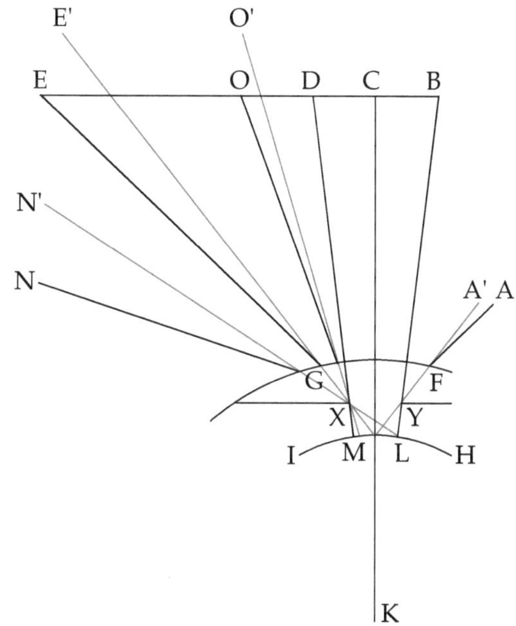

Fig. 5.11 Formation of Della Porta's funnel of radiation

center point K of the eye. Thus, B and D will be seen where they actually are from viewpoints M and L.

Meantime, rays EG and AF reach the cornea at a slant so as to refract through the pupil and into the aqueous humor along GM and FL. According to the reasoning in the previous paragraph, object A will appear to lie where the extension of refracted ray FL intersects the cathetus dropped from center point K of the eye through object A. It will therefore appear somewhere beyond A. The same holds *mutatis mutandis* for object E, which will appear to lie somewhere beyond E along the extension of refracted ray GM. From this analysis, it follows that the actual "cone" of radiation takes the shape of a funnel (*infundibulum*) AFLMGE, whose outer ray couples AFM and EGL are determined by the outer edge of the pupillary opening GF.

Precisely what Della Porta is getting at here is a bit unclear because the diagram misrepresents his analysis somewhat by locating the pupil on the very surface of the cornea rather than just below it—all, perhaps, for the sake of simplicity. If the diagram is redrawn to reflect the pupil's actual location, the funnel takes on a slightly modified shape. Let Fig. 5.11 reflect the same situation as represented in Fig. 5.10, but with the pupillary opening XY located between the cornea GF and the lens HLMI. As in Fig. 5.10, perpendicular rays BL and DM form the outer edge of the cone of radiation centered on K, CK being the cone's axis and therefore the visual axis. As in Fig. 5.10, also, incident ray EG to the side of the cone is refracted on entering the aqueous humor, but this time it skims edge X of the pupil on its way to the lens. The image of E is thus seen along extension XGE' of the refracted ray, so it is seen on the same side of visual axis CK as its object. The same holds for the incident rays from O and N. They too are refracted at the cornea in such a way as to skim edge X of the pupil, the resulting refracted rays extending toward images O' and N', both of which are on the same side of visual axis CK as their generating objects. Consequently, the overall image CO'E'N' will have the same orientation as line COEN, and so will its symmetrical counterpart on the other side of visual axis CK.

Admittedly, this reconstruction of Fig. 5.10 is speculative, as is the resulting interpretation of the analysis in book 4, proposition 9, to which the figure is tied. The virtue of this reconstruction, as well as the interpretation that flows from it, is that it makes Della Porta's model of peripheral vision logically consistent. This it does by providing a clear account of precisely how and where any point in the field of view will be seen according to a specific, and specifically defined, incident and refracted ray-couple. What the reconstruction fails to do is provide any explanation, whether physical, physiological, or psychological, of *why* this specific ray couple yields a proper view of the given point that corresponds to it. Moreover, it assumes that Della Porta made a conscious and concerted effort to ensure his account of vision, in all its aspects, would be as logically consistent and coherent as possible. As we will see in the next section, this assumption is highly questionable.

5.4 The Problem of Diplopia

In the very late sixteenth century, when Della Porta composed his *De Refractione*, the account of diplopia, or double vision, in the third book of Alhacen's *De aspectibus* was pretty much canonical (Alhacen 2001, 247–85 [English], 562–88 [Latin]). The gist of that account is as follows, according to Fig. 5.12. Let C be the center of sight in the left eye, B its counterpart in the right eye, and A an object facing the two eyes. In the left-hand diagram, the axes from both eyes converge on object A, so it

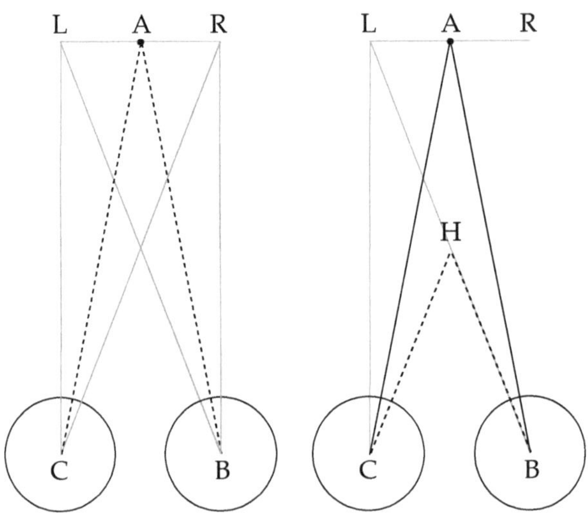

Fig. 5.12 Binocular vision and diplopia according to Alhacen and the Perspectivists

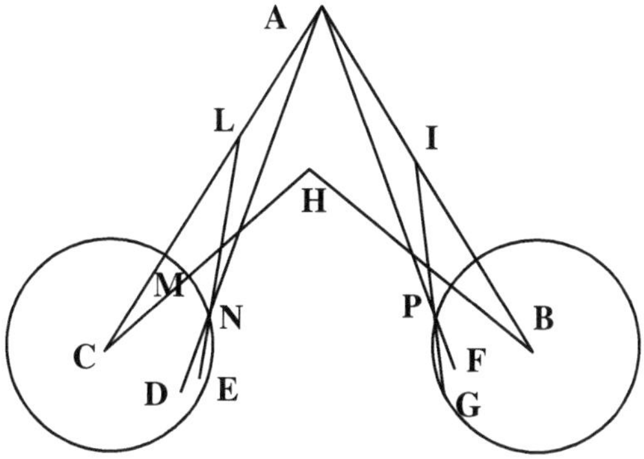

Fig. 5.13 Example of diplopia according to Della Porta

will appear to lie straight ahead from both viewpoints C and B. It will also appear single because the two images radiating along axes CA and BA will strike equivalent, or at least closely corresponding, points on the eye. The rays incident along axes AB and AC thus correspond as perfectly as possible.[17]

If, however, the two axes converge on point H, as in the right-hand diagram, then A will appear double because two images radiate from it along non-corresponding rays to non-corresponding points on the eyes. Accordingly, one image of A will reach the left-hand eye along ray AC to the left of visual axis CH, whereas the other will reach the right-hand eye along ray AB to the right of visual axis BH. It follows, then, that from the perspective of viewpoint C, object A will be seen to the left of H—which appears to lie straight ahead according to the convergence of the two visual axes—whereas the same object A will seen to the right of H from the perspective of viewpoint B.[18]

Clear and reasonable though it is, Della Porta chose to reject this explanation, presumably because it depends on a more expansive cone of radiation than the pupil will allow. With this point apparently in mind, Della Porta turned to his account of peripheral vision for an appropriate explanation, thus reducing diplopia to a special case of peripheral vision. For instance, in Fig. 5.13, which is drawn directly from

---

[17] See esp. Alhacen 2001, 247–53 [English], 563–68 [Latin]. Alhacen's understanding of correspondence in this case is illustrated in the left-hand diagram of Fig. 5.12, where line LAR forms the base of radiative cones LCR and LBR. From viewpoint C, endpoint L of line LAR is seen along CL to the left of visual axis CA, and it is viewed under visual angle LAC. From viewpoint B, meantime, that same endpoint L is seen along BL to the left of visual axis BA, and it is seen under visual angle LBA, which is virtually equal to visual angle LCA. Hence, as Alhacen puts it, rays LC and LB have a "corresponding situation" (*positio consimilis*), which means that the two images radiating along LC and LB will strike the surfaces of both eyes at virtually the same point with respect to where visual axes AC and AB intersect the surface of their respective eyes. The same holds for endpoint R, so it follows that L, R, and all points in between will be seen single, as will the entire line LAR.

[18] Thus, in the right-hand diagram of Fig. 5.12, endpoint L will be seen by viewpoint C along LC and by viewpoint B along LB. Since LC lies to the left of axis CH and forms sizeable visual angle LCH with it, whereas LB lies slightly to the right of axis HB and forms an almost imperceptible visual angle with it, the image of L will strike each eye at non-corresponding points, as will the image of R, which means that the entire line LR will be seen double.

book 6, proposition 3, of the *De Refractione* (Della Porta 1593, 147–48), A represents the object, C and B the left and right eyes, respectively, and BH and CH the visual axes—or *lineae centricae*, as Della Porta designates them.[19]

Now according to the Alhacenian analysis discussed in the previous two paragraphs, object A will appear double because it is seen along non-corresponding rays AC and AB. Della Porta supposes instead that the two images of A reach the eyes along rays AN and AP, thus striking their corneal surfaces obliquely and being refracted by the aqueous humor. It follows, then, that instead of continuing straight along ND, left-hand ray AN will refract along NE, whose extension intersects cathetus CA at point L. From the resulting viewpoint on the lens, i.e., where refracted ray NE strikes its anterior surface, image L will be seen to the left of H along ENL. Likewise, right-hand ray AP will refract along PG after striking P on the right hand eye, so the resulting image I on cathetus BA will be seen to the right of H along GPI.

The problem with this analysis is that it flouts what every optical authority from antiquity to Della Porta's day took to be an undisputed and universal principle of refraction: namely, that in passing obliquely from a rarer to a denser medium, light will be bent toward the normal. In this case, however, the two rays are refracted *away* from the normal and thus diverted to the wrong side. Accordingly, incident ray AN should be refracted to the left of ND, toward C rather than to the right along NE, away from C. The same holds for AP. It, too, should be refracted toward B rather than away from it along PG. As a consequence, if incident rays AN and AP were refracted properly, and if the resulting rays NE and PG were to intersect their respective catheti appropriately, the analysis would be entirely different in detail, even though in certain cases it can lead to the same conclusion about the orientation of the two images.

This point is illustrated in Fig. 5.14, which recapitulates the basic conditions in the right-hand diagram of Fig. 5.12, with A being the object, C and B the centers of the eyes, and CH and BH the visual axes or centric lines converging on point H in front of the object. In conformance with the appropriate principle of refraction, incident ray AN will be refracted toward the normal (and thus toward C) along NE, and when NE is extended, it will intersect the extension of cathetus CA at image location L. Viewed along ENL, therefore, image L will be seen to the left of H, just as was the case according to Della Porta. Likewise, since it is viewed along GFI, image I will still be seen to the right of H, which is where Della Porta located it. The primary difference between this analysis, which is based on a correct understanding of refraction, and the analysis before it, which is based on an incorrect understanding, lies the locations of L and I with respect to A. In Fig. 5.13, image L lies to the left of A, image I lies to the right of A, and both lie in front of it. In Fig. 5.14, on the other hand, although they still bear the same left-right, line-of-sight orientation with

[19] From the vocabulary he uses to describe not just the axial ray but also the outermost and intermediate rays within the visual cone, it is quite clear that Della Porta was following Leon Battista Alberti's description in book 1 of the *De Pictura*. Cf. Alberti 1540, 10–18, esp. 14–17, and Della Porta 1593, 99.

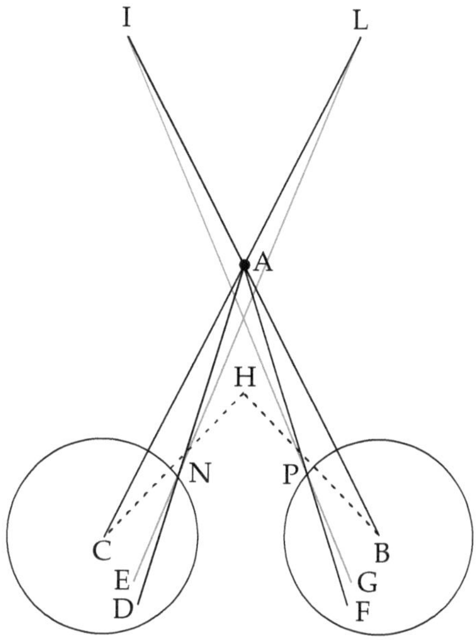

Fig. 5.14 Modified version of Della Porta's example of diplopia

respect to H, the two images shift sides relative to A and are located well beyond it. In addition, depending on where H and A are located relative to one another, it is possible for refracted ray EN to continue to the right rather than the left of H, and vice-versa for refracted FG. In that case, I and L will occupy opposite positions from those illustrated in Figs. 5.13 and 5.14.

There is no way of knowing why, in this particular analysis of diplopia, Della Porta rejected what everyone before him recognized as a cardinal rule of refraction. Nor is this the only occasion in which Della Porta violated that rule, even though he did not violate it consistently.[20] But it is crucial to bear in mind that these "demonstrations" were actually illustrations rather than proofs. Della Porta in fact proved nothing about diplopia; he merely provided a geometrical description, so it is possible that he simply ignored the norms of refraction in order to make his point as clearly and simply as possible. Or it may be that he was driven to violate those norms in order to make his analysis fit the phenomena as closely as possible.

Whatever his motivation, Della Porta played fast and loose with standard optical principles, not just in the analysis of diplopia but at other points in the *De Refractione* as well. Perhaps the most egregious example of this disregard for norms comes in proposition 2 of the third book (Della Porta 1593, 68–71), where he undertakes to show why the eye has to be a globe. One reason, Della Porta suggests, is that if the surfaces of the eye and lens were somehow concave rather than convex, everything

[20] For further instances of violation, see book 6, props. 5–8 (Della Porta 1593, 149–53); for an instance of non-violation, see book 6, prop. 4 (ibid., 148–149); and for an instance in which the rule is both violated and followed in the same theorem, see book 6, prop. 9 (ibid., 153–54).

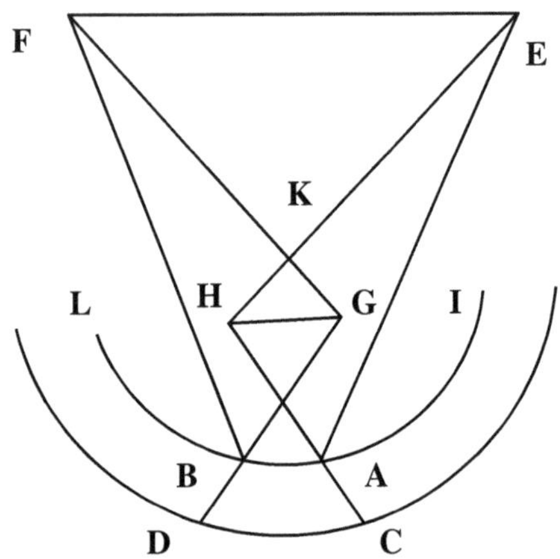

Fig. 5.15 Della Porta's illustration of image-invesion if eye were concave

would appear inverted. His demonstration is based on Fig. 5.15 (Della Porta 1593, 70), where arc LBAI represents the supposed concave outer surface of the cornea, arc DC the concave anterior surface of the lens, and K the center of curvature for both. Let FE be an object facing this eye, and let ray EA from its right-hand endpoint strike outer surface ABAI obliquely at point A. Likewise, let ray FB from the left-hand endpoint strike that same surface obliquely at point B. Both rays will therefore refract into the aqueous humor, EA along AC and FB along BD. Consequently, if refracted rays AC and BD are extended to their respective catheti EKH and FKG, the image of object FE will be HG, whose right hand point G is the image of left-hand point F on the object, and vice versa for point H on the image and point E on the object. The object will therefore appear reversed.

Clear and compelling though this analysis might seem, it suffers from two fatal flaws. The first is that the conclusion is false. Under the conditions illustrated in the figure, the image of object FE will have the same orientation as the object itself. In other words, the image of FE will not be reversed. The second fatal flaw is the source of the first: namely, the supposition that rays EA and FB will refract along lines AC and BD, both of which not only verge *toward* the normal, as they should in entering the denser albugineous humor, but actually pass through and beyond it. In order to make this point clear, let us look briefly at Fig. 5.16, which is based on Fig. 5.15. According to the previous analysis rays EA and FB refract along AC and BD (in gray), respectively, so when those two refracted rays are extended to their respective catheti, EK and FK, they will form endpoints of reversed image HG (also in gray). But when we draw normals KAY and KBX through points of refraction A and B, we see immediately that AC and BD pass beyond, and thus to the wrong side of, their respective normals. In reality, the refracted rays should verge toward, but never meet, the normals, much less pass beyond them. Hence, EA will refract along some line AC' short of normal KAY, and FB will refract along some line BD' short of normal KBX. As a result, when refracted rays AC' and BD' are extended, they will meet their respective catheti EK and FK at H' and G', so the image (in black) defined by endpoints G' and H' has the same orientation as its object FE.

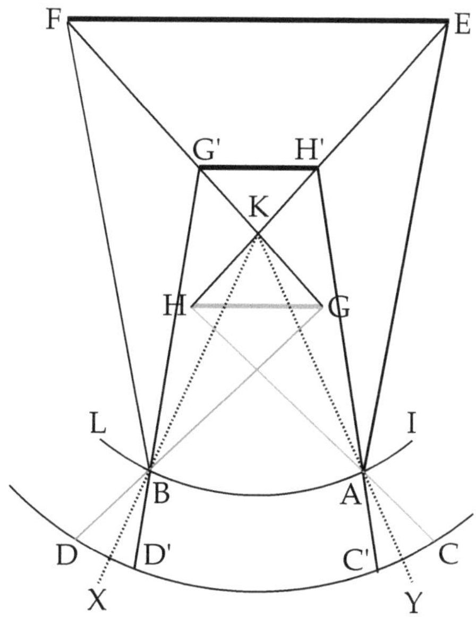

Fig. 5.16 Correction of Della Porta's conclusion about image-inversion if eye were concave

5.5 The Problem with Image Location

As we saw in the previous section, Della Porta went badly astray in accounting for image formation in diplopia because he was apparently convinced that light can sometimes refract away from the normal when passing obliquely into a denser medium. As a result, his demonstrations were often based on improperly determined refracted rays, which in turn defined inappropriate lines of sight. But, image-location depends not just on the lines of sight established by refracted rays. It is also a function of where the cathetus of incidence intersects those rays. Image location, in short, is subject to the so-called cathetus rule.

The cathetus itself in the line dropped orthogonally from a point on an object to the surface of refraction. In the case illustrated in Fig. 5.17, which accompanies book 1, prop. 7 (Della Porta 1593, 15–16), the object I lies underwater, the viewpoint A lies in the air above the water's surface to the side of I, and ray IB strikes the plane surface of the water obliquely at B and is refracted to A along AB. AB thus defines the line of sight, and the image L is located where the extension of refracted ray AB intersects cathetus IL dropped orthogonally to the water's surface. If the refracting surface is curved, then the cathetus dropped from the object will pass through the center of curvature, as in Fig. 5.8, in Sect. 5.3 above, where the cornea and lens of the eye are represented by the outer and inner arcs, the object by B, the incident ray by BC, the refracted ray by CD, the cathetus by FB, and the image by A.

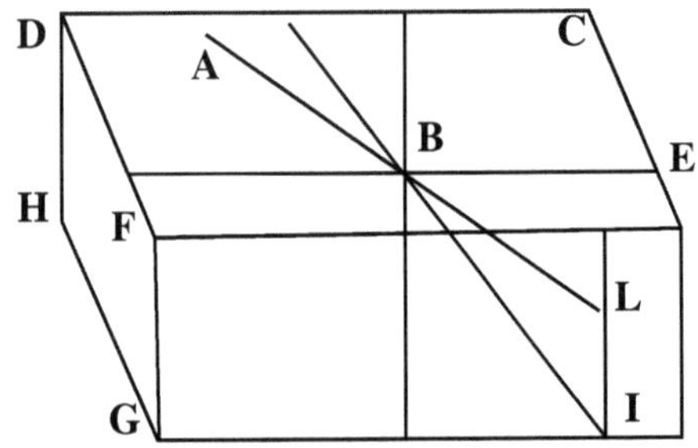

Fig. 5.17 Image-location according to the cathetus rule

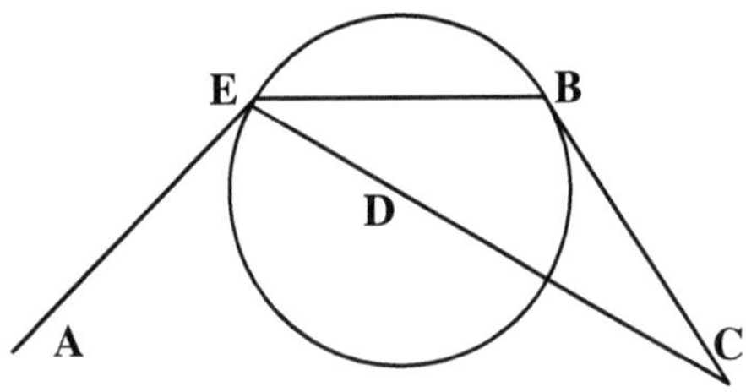

Fig. 5.18 Image-location on the surface of a glass sphere according to Della Porta's analysis

That Della Porta both understood and subscribed to the cathetus rule is evident not only from these two examples, but also from the way he applied it throughout the *De Refractione*. Indeed, as Arianna Borelli points out in a recent study (Borelli 2014, esp. 53–59), Della Porta uses that rule to good effect throughout book 2, where he accounts for various visual effects produced in spherical glass lenses. Included among these effects is the way images can appear to lie on the actual surface of the sphere or even to "hang in the air," and he explains them all according to what Borelli characterizes as a "methodically precise, conceptually clear and empirically accurate" procedure (Borelli 2014, 56).

This procedure, which falls squarely within the classical Greek and medieval optical tradition, is exemplified in book 2, prop. 8, from which Fig. 5.18 is drawn (Della Porta 1593, 49). Della Porta's point in this proposition is to show that "if the cathetus intersects the line [of refraction] conveying the image at the surface of a [glass] sphere, the image wil be seen at that point."[21] In other words, from viewpoin A in Fig. 5.18, object C will be seen at point E on the surface of the glass sphere represented by the circle centered on D. The "proof" rests on the fact that the cathetus dropped from C through the sphere's center of curvature D intersects the refracted ray AE at E, which is the image location and lies on the sphere's surface.

Things begin to go awry, however, when Della Porta moves from point objects to objects of appreciable size. Take book 2, prop. 13 (Della Porta 1593, 53–55), as an example. Della Porta's purpose here is to show that, "when a magnitude lies at the

[21] Catheto secante lineam deferentem formam in superficiem pilae, ibi videbitur simulachrum. Although Della Porta uses crystal (*crystallinus*) and glass (*vitreus*) interchangeably in describing the transparent sphere (*pila*), I have rendered both terms "glass" for the sake of consistency, reserving "crystal" for the eye's lens.

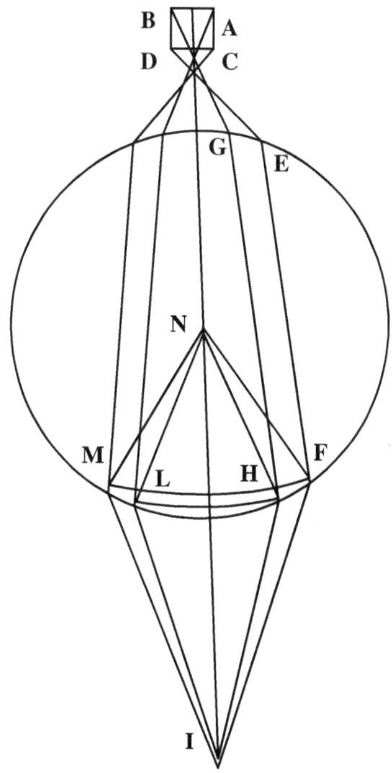

Fig. 5.19 Formation of annular image on glass sphere's surface according to Della Porta

end of the diameter, ... its image will be seen on the surface of a [glass] sphere in a ringlike form."[22] Based on Fig. 5.19, the demonstration is as follows. Let BACD be a relatively small object facing viewpoint I on the other side of the glass sphere centered on N, and let diameter IN pass through the middle of the object. The image of D will strike the sphere's surface at E, then refract into the sphere along EF, and exit the sphere along line of refraction FI, which defines the line of sight along which D's image should be seen. Likewise, the image of B will reach I via rays BG, GH, and HI, this latter defining the line of sight. Since the catheti dropped through center of curvature N intersect HI and FI at H and I, respectively, it follows that those two points constitutes the images of their objects B and D. Hence, if we rotate the sphere about axis NI, points H and F will describe the inner and outer circumferences of a ring.

The flaw in Della Porta's reasoning is that, according to the proper definition of "cathetus," as articulated and applied by Della Porta himself in book 1, prop. 7, neither NF and NH is a cathetus because neither passes through objects B or D. Under the conditions illustrated in Fig. 5.19, in fact, the appropriate catheti dropped from B and D through center point N will intersect their respective refracted rays HI and FI not on the sphere's surface itself but between that surface and viewpoint I. Both points should thus appear to "hang in the air" in front of I. This is

[22] Magnitudine tangente finem diametri, ... imago videbitur in pilae superficie quemadmodum annularis.

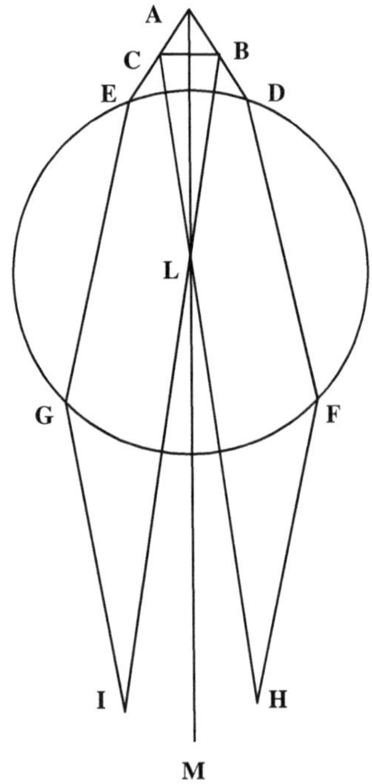

Fig. 5.20 Seeing a double image of a single object through a glass sphere according to Della Porta

actually not the case, though. If the object is small enough, and if it lies in a straight line with the viewpoint and the sphere's center, it will form an image that seems to spread over the front surface of the sphere in a ringlike pattern. Faced with this empirical fact, therefore, Della Porta apparently chose to account for it by "bending" the cathetus rule to fit the phenomenon.

An even more startling case of such "bending" can be found in book 2, prop. 15 (Della Porta 1593, 56), where Della Porta undertakes to show that more than one image of a single object can be seen in a glass sphere. By way of demonstrating this point, Della Porta invites us to suppose that I and H in Fig. 5.20 represent the left and right eyes, respectively, that both of them face a glass sphere centered on L, and that A represents a single object viewed through the sphere. According to the analysis based on this figure, the image of A reaches left-hand viewpoint via incident ray AE, refracted ray EG, and re-refracted ray GI. By the same token, the image of A reaches H through a similar succession of incident, refracted and re-refracted rays. In order to locate the two images of A, Della Porta has us extend line HL to point C and line IL to point B. C will therefore be the image of A seen from viewpoint H, whereas B will be the image of A seen from viewpoint I.

Bizarre in the extreme, Della Porta's reasoning in this theorem suffers from three truly bewildering logical inconsistencies. For one thing, HL and IL are not catheti in the proper sense of the term. In fact, there is only one cathetus represented in the diagram, i.e., the line passing from object A through center point L, which does not figure at all in Della Porta's analysis. For another thing, even granting that HL and

LI can be thought of as catheti according to Della Porta's rather flexible interpretation of the term, images C and B are formed by them on the wrong rays, AE and AD being incident rather than refracted rays. And for yet another, the image should be seen along refracted lines of sight IG and HF, not along supposed catheti IL and HL. Consequently, unlike the previous example, which suffered from a single logical flaw, this one suffers from three, one stemming from an improper designation of "cathetus," another from an improper application of the cathetus rule itself, and still another from an improper designation of the line of sight.

5.6 Conclusion

Cursory though it is, this overview of Della Porta's account of sight in the *De Refractione* is certainly adequate to show, first, that his theory of vision in no way anticipates that of Kepler, with its basis in retinal imaging. Far from pointing toward the Keplerian future, in fact, Della Porta's theory is in many ways still rooted in the classical and medieval past, particularly the Perspectivist past. Yet it is also evident that Della Porta was not mired in that past. Quite the contrary, he was willing to part ways with his classical and medieval antecedents at several reprises and, in the process, forge off in analytic directions that were at best problematic and at worst irrational. Just how problematic the results could be is manifest in his accounts of peripheral vision, diplopia, and image location.

The resulting lapses and logical inconsistencies—some of them fairly glaring—are puzzling because they suggest that Della Porta was either ignorant of certain optical principles or unable to see the flaws in his own reasoning. It is difficult to credit either of these suggestions. Even a cursory reading of the *De Refractione* leaves no doubt that Della Porta had a firm grasp not only on the fundamental principles of optics, as understood in his day, but also on how those principles were applied by a wide range of optical thinkers before him. As a student of optics, in fact, he was extraordinarily erudite, availing himself not just of printed sources but also of manuscripts.[23]

[23] Ranging from classical to medieval and contemporary works, Della Porta's sources include several genres. Among those that fall into the category of natural philosophy or "science," for instance, we can include Aristotle; Euclid; Hero of Alexandria; Ptolemy; Galen; Alexander of Aphrodisias; Olympiodorus; Avicenna; al-Kindī; Albertus Magnus; Johannes de Sacrobosco; Alhacen and his perspectivist followers Roger Bacon, Witelo, and John Pecham; Andreas Vesalius; Girolamo Cardano; and Giulio Cesare Scaligero. In addition, Della Porta used such quasi-scientific sources as Pliny's *Natural History*, Aelian's *De varia historia*, Aratus's *Phaenomena*, Macrobius's *Saturnalia*, Solinus's *De rerum mirabilium*, and Seneca's *Naturales quaestiones*. He also makes explicit reference to Vitruvius's *De architectura*, Leon Battista Alberti's *De pictura*, and the 1532 Latin version of Albrecht Dürer's *Underweysung der Messung*, and he even consulted such literary sources as Xenophon's *Anabasis*, Plutarch's *Moralia*, Cicero's *Academic Questions*, Virgil's *Aeneid*, Suetonius's *Histories*, and Aulus Gellius's *Attic Nights*.

Furthermore, Della Porta's avidity as a reader of optical sources was fully matched by his understanding of them, a point borne out by his trechant critique of technical details in many of those sources, particularly the Perspectivist ones.[24] Della Porta did more than simply parrot his sources; he subjected them to critical scrutiny and in the process offered significant modifications to them. His discussion of pupillary contraction and dilation in the third book of the *De Refractione* is a clear case in point; and so is his effort to explain why the pupil is subject to such size-variations according to light intensity. No less original is his all-but-pioneering analysis of the focusing properties of concave spherical mirrors and lenses or, for that matter, the effort to account for diplopia according to his model of peripheral vision.

Why, then, did Della Porta go so badly astray at certain points in his analysis? One ostensible reason is that he needed to adjust the rules (or least thought he did) in order to account for the actual, observed phenomena. That, for instance, would explain why his definition of "cathetus" was so flexible, and why the resulting derivation of image location was so inconsistent. But to assume that Della Porta bent or broke the rules in order to accommodate the appearances opens the question of why he felt justified in following such a procedure. Did he regard all these rules as merely provisional and thus adaptable to particular circumstances?

Perhaps he did, but it is impossible to answer this question definitively because Della Porta never discussed his theoretical or methodological presuppositions at a general, philosophical level. The best we can do is suggest factors, beyond mere analytic convenience, that may have driven him to bend or break the rules as he did. One that deserves serious consideration is his marked preference for physical over physiological or psychological causation in explaining visual phenomena. This preference is reflected in his effort to naturalize the visual process by rejecting such occult causal mechanisms as visual spirits in the optic system or animal spirits in the brain. "I wonder," Della Porta marvels, "at how frightfully insane it is to imagine that… spirits … convey phantasms through the hollows of the optic nerves and that natural spirits convey intellectual images [through the brain]" (143).

By rejecting the physiology of cerebral and visual spirits, Della Porta also rejected the scholastic model of visual sensation and perception based on them. No longer could he appeal to the physiologically and psychologically determined center of sight that the Perspectivist optical theorists had adduced as the cardinal reference point for visual analysis. Nor could he appeal to the Perspectivist cone of visual perception because, by repudiating virtual perpendicular radiation he was compelled to transform that cone into a funnel. With these options foreclosed, he had to temporize in order to account for the various phenomena associated with sight, particularly those of spatial perception. Still wedded to the Perspectivist assumption of lenticular sensitivity, though, Della Porta was forced to accommodate his account to that assumption. The result, as we have seen, was neither a new, overarching theory nor a full-scale revamping of the old one.

[24] See, e.g., Della Porta 1593, 17, 21, 63, and 79, where he takes Witelo, Alhacen, Bacon, and Pecham, respectively to task for various analytic errors.

References

Alberti, Leon Battista. 1540. *De Pictura*. Basel: Westheimer.

Alhacen. 2001. *Alhacen's Theory of Visual Perception: A Critical edition, with English Translation, Introduction, and Commentary, of the First Three Books of Alhacen's* De aspectibus, *the Medieval Latin Version of Ibn al-Haytham's* Kitab al-Manazir. Ed. and Trans., A. Mark Smith. Philadelphia: American Philosophical Society.

———. 2010. *Alhacen on Refraction: A Critical Edition, with English Translation, Introduction, and Commentary, of Book 7 of Alhacen's* De aspectibus. Ed. and Trans., A. Mark Smith. Philadelphia: American Philosophical Society.

Bacon. 1983. *Roger Bacon's Philosophy of Nature: A Critical Edition, with English Translation, Introduction, and Notes, of* De multiplicatione specierum *and* De speculis comburentibus. Ed. and Trans. David C. Lindberg. Oxford: Clarendon Press.

———. 1996. *Roger Bacon and the Origins of Perspectiva in the Middle Ages: A Critical Edition and English Translation of Bacon's* Perspectiva *with Introduction, and Notes*. Ed. and Trans. David C. Lindberg. Oxford: Clarendon Press.

Borelli, Arianna. 2014. Thinking with optical objects: glass spheres, lenses and refraction in Giovan Battista Della Porta's optical writings. *Journal of Early Modern Studies* 3: 39–61.

Della Porta, Giambattista. 1589. *Magiae Naturalis Libri XX*. Naples: Apud Horatium Salvianum.

———. 1593. *De Refractione Optices Parte Libri Novem*. Naples: Apud Io. Iacobum Carlinum et Antonium Pacem.

Donahue, William. 2000. *Johannes Kepler: Optics*. Santa Fe, NM: Green Lion Press.

Kepler, Johannes. 1604. *Ad Vitellionem Paralipomena Quibus Astronomiae Pars Optica Traditur*. Frankfurt: Apud Claudium Marnium et Haeredes Iohannis Aubrii.

Lindberg, David C. 1975. *A Catalogue of Medieval and Renaissance Optical Manuscripts*. Toronto: University of Toronto Press.

Smith, A. Mark. 2015. *From Sight to Light: The Passage from Ancient to Modern Optics*. Chicago: University of Chicago Press.

Vesalius, Andreas. 1568. *De Humani Corporis Fabrica Libri Septem*. Venice: Apud Franciscum Senensem et Iohannem Criegher Germanum.

Chapter 6
Magi from the North: Instruments of Fire and Light in the Early Seventeenth Century

Fokko Jan Dijksterhuis

Τοῦ δὲ Ἰησοῦ γεννηθέντος ἐν Βηθλέεμ τῆς Ἰουδαίας ἐν ἡμέραις Ἡρῴδου τοῦ βασιλέως, ἰδοὺ μάγοι ἀπὸ ἀνατολῶν παρεγένοντο εἰς Ἰεροσόλυμα (Matthew 2:1)

When the news of the telescope began to spread through Europe, the old Della Porta once again took up his pen. He wrote an exposition on the telescope, explaining its effects by elaborating on his earlier accounts of refraction and lenses. With this exposition Della Porta not only gave evidence of his intellectual command of the instrument, he also laid claim to its authorship: ' and it pleases me that the idea of the telescope in a tube has been mine; ...'[1] It was a particular way of making a claim. Della Porta did so by mathematical means, showing that he understood the instrument and thus was the intellectual author, even if he had not been the first to actually build it. Della Porta's claim was to no avail: *De Telescopio* was not published, the manuscript did not circulate, and the author died not long after its conception. Whether the claim was justified, is not relevant here; fact is that Della Porta was not considered to be an inventor of the telescope at that time.

A text like *De Telescopio* can be read on several levels. On an earlier occasion I interpreted it as a theoretical exposé, assessing its analytical cogency.[2] In *De Telescopio* Della Porta extended his account of refraction in transparent spheres from *De Refractione* to configurations of lenses in order to explain the workings of the telescope. He quite inventively applied the cathetus to a succession of refracting surfaces in order to explain the magnified image produced by a (Galilean) telescope. From the perspective of Kepler's new theory of image formation 'Della Porta's theory of lenses was fraught with difficulties and mathematically it was riddled with ambiguities', I wrote 15 years ago. Since then it has become clear that a Keplerian perspective is not necessarily appropriate to interpret Della Porta's optics. In particular Arianna Borrelli's recent reading of Della Porta has made clear that this is

[1] Ronchi 1954, 56 and 34. "They know nothing of perspective."

[2] Dijksterhuis 2004, 33–35.

F.J. Dijksterhuis (✉)
University of Twente, Enschede, Netherlands
e-mail: f.j.dijksterhuis@utwente.nl

A. Borrelli et al. (eds.), *The Optics of Giambattista Della Porta (ca. 1535–1615): A Reassessment*, Archimedes 44, DOI 10.1007/978-3-319-50215-1_6

too one-sided and limited.[3] Della Porta explained the properties of lenses in terms of the effects of artefacts on the images as they are perceived. The meticulous way in which Borrelli shows how Della Porta extended his understanding of spheres to the effects of lenses has fundamentally changed my understanding of *De Telescopio* and of the telescope as an instrument of natural magic.

Inspired by this I want to approach optics in early modern natural magic with fresh eyes and reconsider categories of optical instruments. I will do so by looking at practices of natural magic rooted in the Low Countries. The lead will be a letter that was induced by the news of the telescope in the same way as Della Porta's exposé. It claimed Dutch authorship of the telescope and although the claim was justified, of interest here is the way it puts the telescope in a setting of natural magic in the same way as Della Porta does. I will use this letter to reconsider the character of the telescope and other early modern instruments of optics. I want to broaden the term 'optics' beyond the usual dioptrics, to a more general sense of controlling and manipulating light, sight and perception. A detour to the Low Countries brings us to Cornelis Drebbel, another resourceful inventor of optical instruments and in many ways comparable to Della Porta. In addition to historicizing the concept of optical instrument, I will use the northern counterpart to reflect upon the epistemic features of natural magic. Finally I will briefly discuss the reception of Della Porta in the Low Countries. There was a prominent tradition natural magic in the North in which the work of Della Porta also found a modest place.

6.1 A Telescope in the Lab

Not long after *Sidereus Nuncius* reached the Dutch Republic, a small book was published in the university town of Franeker, in the province of Friesland. *Biolychnium, seu Lucerna, cum vita ejus Cui Accensa est Mystice vivens iugiter*.[4] The treatise discussed a lamp that burned on blood and in so doing revealed the vitality of the owner of the blood. The lamp of life had been invented by Johann Ernst Burggrav († 1629). An earlier edition of the treatise had been published the year before in Leiden, but Burggrav had extended it considerably.[5] The basic idea was that our blood attracts the vital powers of the heavens and thus guides our health. The lamp burned the blood in such a way that the presence of these powers was made visible. The lamp reflected, in other words, universal vitality. *Biolychnium* was much discussed in the early modern period and went through several editions in the seventeenth century.[6] In this way the preliminary matter of the new edition also became widespread, and this is of interest here. Besides preface, panegyric, and

[3] Borrelli 2014. See also Dupré 2005, 2006.

[4] Burggrav 1611.

[5] Burggrav 1610.

[6] Keller 2008, 111 footnote 271.

so on, the book included a long letter to the author, written by Marcellus Vranckheim (1587–1644) from Padua, dated 30 November 1609.

Vranckheim discussed the telescope and other Dutch feats:

> Jacob Metius (the brother of Adriaan Metius, a celebrity because of his remarkable achievements in the mathematical sciences), who invented the spyglass, with which he can measure a tower or any other object from a distance of three Dutch miles, as if he stands right before and eye to eye with it, and with which he can observe England clearly from his beach. He also gives other observations of the surface of the moon, the Milky Way, stars that astronomers have thus far referred to as nebula, and about other stars that are wandering around Jupiter, an unprecedented novelty with regard to earlier generations.[7]

In an overly humanistic style, larded with Greek quotations, Vranckheim continued with disputing the priority of Galileo regarding the telescopic discoveries of *Sidereus Nuncius*. Vranckheim actually had been in Padua when Galileo had started his telescopic observations. Still, the letter was demonstrably antedated and the coincidence of Metius's alleged discoveries with those of Galileo is suspicious to say the least. We have little to no information about the observations Metius made with his instruments, if he ever used them at all. We do know, however, that Vranckheim's claim that Jacob Metius had invented the telescope was right.[8] The Vranckheim letter was reprinted several times and widely read; it was a major source for the attribution to Metius, for example by Descartes in *La Dioptrique*. The idiosyncratic letter of Vranckheim is interesting; not so much for its factual claims regarding inventions and discoveries, but for the context in which it places the early telescope and Metius' authorship.

Marcellus Vranckheim was a young man from Zutphen. After a spell at Leiden, late 1608 he had gone on a *peregrinatio academia* that brought him to Marburg, Basel, and eventually in Venice and Padua in late 1609.[9] In Marburg he had acquainted Burggrav, a well-travelled scholar and medicus. From Marburg they went their own ways. While Vranckheim travelled south to Italy, Burggrav went to England and other places. They met again in Franeker. After his return to the Republic Vranckheim matriculated at the university in September 1610. In Franeker

---

[7] Burggrav 1611, 53–54. Transcription and translation Dijkstra 2012, 145. 'Sed ecce tibi alterum huic concivem! Iacobus Metius est (frater Adriani Metij, Viri ob singularem in Scientijs Mathematicis praestantiam Clarissimi) qui Perspicillum invenit, quo turrim vel corpus aliud quodlibet tribus milliaribus Hollandicis dissitum, velut pede collato, & ad oculum dimetitur exactis simè, & Angliam è littore suo clarè prospicit, & alia de Lunaeglobo; de Galaxia; de Stellis, quas Nebulosas hactenus dixerunt Astronomi; de alijs circa Iovem erraticis prodit inaudita veterum aevo novitate. 'Quam penè, τάς μηχανάς μετά τον πόλεμον mi Galilee, [Galileus Galileus Matheseos in illustri Patavina Professor] & perspicillum illud tuum cum Observationibus? Et Sidereus Nuncius tuus Callipides erat, nisi inter caesa & porrecta, quod aiunt, nescio quid additum fuisset operae homini Batavo, & adhuc imberbi, qui citiùs Instrumentum tale invenerit, quàm ejus à se reperti famam sparserit, adeo ut rumore de hoc divulgato, ad consimilis Organi inventionem te devenisse confitearis ipsemet, cujus beneficio Observationes εκ τών υπερβερεταίον illas prodidisti in Lunae facie; fixis innumeris; lacteo circulo; stellis nebulosis; & quatuor Planetis, eorundemque circa Iovem periodis. Age verò, dum alteram tibi muralem cedimùs, ne Isthmum Sinapi bibas.'

[8] Zuidervaart 2010, 15–16.

[9] Dijkstra 2012, 142–143.

the two collaborated with the renowned professor of mathematics, Adriaan Metius, the brother of Jacob.

Adriaan Metius was an active writer and purposive marketer. He wrote several textbooks on mathematics that went through numerous editions, making him the widest circulated and best read mathematician of the seventeenth century.[10] In his books Adriaan actively promoted the achievements of his family. Besides the achievements of his father – who was the main engineer in the States Army and a mathematician of note – he would repeatedly mention Jacob's telescope in his textbooks on astronomy. The letter of Vranckheim was a timely exhibit of the priority of his brother and Metius was probably instrumental in the publication of the second edition of Burggrav's *Biolychnium*. The *Arithmeticae* of Metius was published at the same press and Vranckheim had written a carmen for it.[11] The Vranckheim letter fitted Adriaan's ambitions to enhance the reputation of his kin. But why would a renowned mathematician use a chymical treatise as the vehicle for the claim? Part of the answer is that Adriaan Metius had a profound interest in medicine and alchemy. He closely collaborated with the resident of the Franeker castle, Carolus Sternsee. They equipped an alchemical laboratory at the castle where they performed experiments, allegedly aimed at curing Sternsee.[12] Burggrav referred to this laboratory in his preface to *Biolychnium* and probably had come to Franeker for the alchemical activities in the first place.

The book of Burggrav was an excellent place for Vranckheim's letter. It placed the lamp of life among a string of products of Dutch ingenuity. It may seem a strange mix of artifacts to modern eyes but from the perspective of early seventeenth-century natural magic it was not. Prominent were a couple of instruments of Cornelius Drebbel, an engineer working in London in the patronage of king James. Vranckheim elaborately described Drebbel's inventions and particularly praised his perpetuum mobile and light organ. Vranckheim based his account on the first-hand experiences that Burggrav had related to him earlier.[13] There was an important feature Vranckheim's Dutch virtuosi had in common: they all were from the Holland town Alkmaar. The Metii were from Alkmaar, where the father's defense of the town in 1573 had been a turning point in the Dutch Revolt. Drebbel also came from Alkmaar, as were other notable Dutchmen of the time like the astronomer and publisher Willem Blaeu (1571–1638) and the maritime explorer Frederik de Houtman (1571–1627). Vranckheim's letter thus also served to promote the reputation of the home town of his protagonists. The beach that Vranckheim referred to was the North-Sea shore of Holland at Egmond, some 10 km from Alkmaar. Of course one could not see England across the North Sea, not even with the telescope of Jacob Metius. But that is not the point; what matters is that the Alkmaar-Metius circle laid claim to the invention of the telescope and did so successfully – for the time being.

[10] Dijkstra 2012, 89–95.

[11] Dijkstra 2012, 146.

[12] Dijkstra 2012, 152.

[13] Keller 2008, 153–154; 390–392; Keller 2010, 51–52.

The connection of mathematics and chymistry was not specific for Metius. It was quite typical of the circles the Vranckheim letter was set in and in which Metius had circulated during his formative years. He had studied in Franeker, one of the schools established by the Nassau family of the Dutch Stadtholders. He had spent time with Tycho in Hven and taught at several protestant German universities including Marburg. In all these places interest in chymistry was common and linked to mathematical pursuits. The professor of mathematics in Marburg during Metius' sojourn there, was Johannes Hartmann (1568–1631). Ten years later, in April 1609, Moritz the learned of Kassel appointed Hartmann as professor *chymiatrie* at the new Collegium Chymicum.[14] During his stay in Marburg, Burggrav resided with the same Johannes Hartmann. Hartmann may well have referred Burggrav and Vranckheim to Metius in Franeker. The main point is that the combination of the letter of Vranckheim, its publication in Burggrav's book, and the place of publication in Franeker, it all reflects the Lower-German milieu of these circles.

The content and context of Vranckheim's letter and the claim by the Metius circle place the telescope in a definite chymical context. This is less peculiar than it may seem. The telescope is usually regarded as an astronomical instrument but we should bear in mind that it has become astronomical only gradually. Even the term astronomical requires qualification when speaking of the early decades of the history of the telescope. In astronomy the telescope was primarily used and understood as a philosophical instrument; used for observing and inquiring into the physical aspects of heavenly bodies. The telescope of *Sidereus Nuncius* is an instrument of cosmology. It became a quantitative instrument only much later, around 1670, and even then the telescope's feasibility as a measuring instrument was contested by someone like Hevelius.[15] With the geometrical analysis of *Dioptrice* Johannes Kepler may have intended to make the telescope fit for positional astronomy, it did not work out immediately. During the first half century, the telescope was an instrument of image making rather than measuring positions.

The episode sketched above mixes up our disciplinary categories. The letter and context of Vranckheim's letter assembles a range of instruments, ideas, and men in which mathematics, magic, chymistry overlap and coincide. This invites us to reconsider our common understanding of the telescope as a mathematical instrument. We can read it not just as an instrument that creates images but in a more general sense one that manipulates light and sight. In this way the telescope naturally fits the Portean category of natural magic; and it did: the telescope was generally connected to mirrors, cameras, and so on.[16] In the company of Burggrav's lamp of life that transforms blood in such a way to make visible the vital element, Jacob's spyglass can be said to transform the visual perception of the world to bring things

[14] Usually regarded as having been the first university chair in alchemy – although this is a rather presentist term. See for example Salloch 2006, 31–39.

[15] Dijksterhuis 2004, 42–46. See also: McKeon 1986; Winkler & Van Helden 1993.

[16] Reeves 2008.

from afar nearby.[17] We can read Burggrav's instrument as an optical instrument as well; not confined to some category of chymistry, but an instrument that deals with light and vision. It adds a chymical element by the substantial conception of light that can be affected in controlled reactions. In this perspective the Franeker telescope was not misplaced in the medical and chymical laboratory of Sternsee: perhaps it could bring to light some vital element as well. The letter of Vranckheim places both instruments alongside each other, and so the ingenuity of Burggrav and Jacob Metius. In this way it enriches our understanding of 'optical' to a general sense of instruments to create and manipulate light and images.

6.2 The Magical Microscope

Some ten years after the annunciation of the telescope, the microscope was received in this same constellation of instruments of natural magic. In London, Cornelis Drebbel entertained a 'standing spyglass' that brought tiny objects and creatures to life size. At the time of the Vranckheim letter he had been at the start of his virtuoso career and after a brief sojourn at the Prague court he had returned to London where he further developed his instruments. Another Dutchman was also in town: Constantijn Huygens (1596–1687) who participated in a diplomatic mission of the Republic in preparation for his career as secretary to the Stadtholder. In the biographical sketch of his youth that he wrote ten years later as a kind of memento for his wife, Huygens described Drebbel's instrument:

> Not only originating from his hand but also conceived by his remarkable brain is an upright telescope, if I may call it so, fitted with two lenses that are both convex. The lowest of these, closest to the object to be observed, is the size of half a nail of the little finger. Even if Drebbel has achieved nothing else in his life, he would have acquired immortal fame with this miraculous little tube. Namely, particles that we used to consider as atoms because they are completely hidden from human sight, the observing eye gets clearly visible before it, so that people who do not know the instrument first complain that they see nothing, because they see things they have never seen. But soon they scream enthusiastically that their eyes see the most incredible things. It is truly as if one stand before a new stage of nature, as if you are on a different world.[18]

[17] Zuidervaart 2010, 11 footnote 11. 'seekere conste … daer mede men seer verre alle dingen can sien al oft die naer bij waeren bij middel van gesichten van glasen, …'

[18] Worp 1897, 119–120. 'Ab eiusdem non manu solâ sed prodigioso ingenio est perspicillum, ut sic dicam, statarium duobus vitris instructum, quorum convexum utrumque, alterum, quod inferius et obiecto proximum est, amplitudine auricularis digiti medium unguem vix adaequat. Hoc mirabili tubulo, ut nihil omni vitâ aliud praestitisset, nominis immortalitatem Drebbelius non dubie promeruit. Corpora nempe, quorum inter atomos hactenus aestimatio fuit, omnem humanam aciem longe fugientia, inspectanti oculo tam distincte obiecit, ut, cum maxime vident imperiti, quae nunquam videre, nihil se videre questi primo, mox incredibilia oculis usurpare clamitent.' My translation is based on the Dutch translation in Huygens 1987.

Early experiences with Drebbel's microscope were of similarly carnal character as Burggrav's lamp of light. It brought to view the worms in the cheese.[19] Rather than inviting a menocchian contemplation of the make-up of the world the experience caused revulsion over this secret of everyday nutrients. For Constantijn it opened entirely new possibilities for artists. He had suggested to the painter Jacob de Gheyn (c. 1565–1629) to employ Drebbel's microscope to depict this uncharted realms and publish a collection of engravings *Novi Orbis*.[20] He continued:

> Therefore nothing will incite us more strongly to worship the eternal wisdom and power of the Creator than entering this treasury of Nature. There, exactly in the smallest and tiniest, we will be confronted with the same devotion of the Architect and everywhere we will come upon the same unspeakable Majesty.[21]

Constantijn expected that with new lenses the tiniest objects could be magnified ad infinitum. According to him bodies are infinitely divisible and all parts have the same properties as the whole.[22]

> I remember that I had very interesting conversations about these matters with Drebbel, who always visited me when I was in London.[23]

Constantijn probably had met Drebbel on his first journey to England in 1618–1619. He mentioned him for the first time in his correspondence during his second stay. Early 1622 he wrote his parents how they had talked about glasses and the English imagining theirs to be better.[24] Constantijn had become a great admirer of Drebbel, whom he regarded together with Francis Bacon as the most significant criticaster of the empty ideas, doctrines and axioms of Antiquity.[25] Unlike Bacon, Drebbel was a pleasant person of modest descent, and he had come miraculously far

---

[19] Constantijn phrases it quite general, but Peiresc around the same time literally viewed cheesemites through it, see below.

[20] Worp, 1897, 120. 'Revera enim istud novo in theatro naturae, alio in terrarum orbe versari est et, si Geinio patri diuturnior vitae usus obtigisset, aggressurum fuisse credo, quo impellere hominem non invitum coeperam, minutissima quaeque rerum et insectorum delicatiore penicillo exprimere compilatisque in libellum, cuius aeri exemplaria incidi potuissent, Novi Orbis vocabulum imponere.'

[21] Worp 1897, 120. 'Infinitam Creatoris Dei sapientiam ac potentiam venerari nullâ re magis adigamur, quam si, satiati obviis cuique hactenus naturae miraculis, quorum, ut fit, frequenti usu ac familiaritate stupor intepuit, in alterum hunc naturae thesaurum immissi, in minimis quibusque ac despectissimis eandem opificis industriam, parem ubique et ineffabilem maiestatem offendamus.' Huygens added in the margin a citation from Aristotle, *De partibus animalium* 1, 5.

[22] Worp 1897, 120–121.

[23] Worp 1897, 121. 'Cum Drebbelio frequenter, quoties Londini essem, ad me visente, memini hac de re praestantissimos sermones fuisse, ...'

[24] Huygens 1911, 76 (letter 120, January 1622). 'J'ay parlé van de brillen mesmes avec Drebbel; il se rit de quoy on s'imagine qu'en Angleterre se fassent les meilleures, ...'

[25] Worp 1897, 112. 'Veterum, quae dixi, inanium notionum, theorematum, axiomatum censores praestantissimos duos aetate meâ suspexi, Franciscum Baconium, Angliae non ita pridem cancellarium, et disparis loci, non ingenii, Cornelium Drebbelium Batavum.'

in physics.[26] Constantijn listed his inventions, lauded his ingenuity, and praised the way this 'Daedalus' combined hand and mind. He was not uncritical of Drebbel's claims and ideas, though, and gave a remarkably balanced report of his feats. He put Drebbel's claims into perspective but he did not agree with people who called him a mere visionary who never lived up to expectations.[27] The microscope did and it made a lasting impression on Constantijn who would bring one home to entertain his family and friends.[28] Constantijn probably first saw Drebbel's microscope during his first stay in England. His companion Willem Boreel later wrote that he had seen it in 1619.[29] In March 1622 Constantijn listed his debts to his parents, saying that the 'lunette' of Drebbel had eaten 40 florins.[30] This was probably a Drebbel microscope – elsewhere the instrument was also referred to as 'lunette' – and the one that allegedly ended up in Constantijn's The Hague home.[31]

Around the same time two of Drebbel's sons-in-law, Abraham and Gilles Kuffler, demonstrated his instruments on the continent. On Tuesday 22 May 1622, the polymath Nicolas-Claude Fabri de Peiresc drew up a lengthy report of a demonstration in the petit Luxembourg in Paris.[32] The 'lunette' of 'Drebellius' consisted of a three-piece tube of some 10 in. in length and one inch in diameter. The lid on one end had a tiny hole behind which a small glass sphere was mounted. The tube on the other end was about a third inch in diameter, holding a lens with a flat side facing the inner lens and the convex side turned to the object on the outside. The convex face was covered with brass, leaving a tiny opening for observation. The demonstrators purported that the glass of the objective lens was of special make, enhancing the clarity of view. Peiresc noted that the image remained remarkable clear and sharp when the instrument was adjusted. The tube was mounted on an adjustable brass tripod. Through the instrument they saw numerous marvelous things, such as the build and color of individual cheesemites, a flea (as well as the fabric of the blotting paper on which it was mounted) that resembled tiny shrimps or crayfish the size of an elver, its egg like a chicken's kidney, and so on with fruit flies, spiders, etc. Like Constantijn they were impressed by the perfection of this tiny life and the way it replicated the structures of the visible world.

---

[26] Worp 1897, 116. 'De Drebbelio, quem cum Baconio copulavi, parcior sermo exit; apposita nempe hac soil luna in physicam praecipue attendebam, qua hunc de plebe Batavum borealem, Alcmariae civem, minim in modem valuisse oculatus testis assero, multa familiaritate hominem perspeetum habens eidemque perspectus.'

[27] Worp 1897, 116. 'Cavillati aliqui cum Jacobo Rege sunt, vix operae quicquam edidisse perpetuum inventorem, cuius utilitate impensa rependeretur'.

[28] Jorink, 2010, 181–184.

[29] Borel 1655, 36. This, however, is not an entirely reliable source as it is also contained the infamous but persistent claim that Sacharias Jansen had invented the telescope. Boreel was also in London in 1622.

[30] Huygens 1911, 91 (letter 141, March 1622). "La lunette de Drebbel en a mangé quarante".

[31] Peiresc. Ruestow 1996 7–11.

[32] Humbert 1951 has made a full transcription. He says that the manuscript is in Peiresc's hand. Jaeger 1922, 133–135 made a full transcription but did not publish the observations.

The Peiresc report is verbal, but in 1631 Isaac Beeckman in Dordrecht made a sketch of a 'perspectivum dr^{j}' that fits the description.[33] The context is illuminating: Beeckman added it to a copy he made of a 1613 letter of Drebbel to King James. This was the moment he returned to London, after his sojourn at the Prague court. Apparently to reapply for patronage Drebbel presented his various inventions. Among these were telescopes with unrivalled properties but not a microscope. This strongly suggests that Drebbel developed his microscope after his return in London, somewhere between 1613 and 1618. Whether Drebbel has actually 'invented' the microscope has been a matter of debate but does not need to concern us here.[34]

The microscope was but one of Drebbel's optical devices. In the course of his career he developed a line of instruments that created images and perceptions in various kinds of ways. With a common trope from the early days of the telescope he told James that he could make telescopes with which letters could be read at 7 miles. He did not claim to have invented the instrument, but he said that his lenses were extraordinary and could not be replicated.[35] Drebbel had first heard of the telescope shortly after its annunciation. Around the time Vranckheim was promoting Jacob Metius's authorship, Drebbel inquired with his acquaintances in his home town after the new invention. 'You have until now remained silent over the 'far seeing' found by the son of Mr. Adriaen Thonissen. Please let me know what he has done. In these I have also found excellent things that seem so incredible and are estimated to be magic'.[36] Drebbel then explained how he would stand in a room in front of an audience and magically change his appearance: first the color and fabric of his clothes, and then taking on the shape of tree, animals and so on. Although the letter is not very explicit, it probably was an early version of the camera obscura that Constantijn Huygens described in critical detail later. 'It's an instrument of simple make, where one can project in a carefully closed room objects that are held outside in front of it right out in the sun.' Drebbel fit a lens in the opening, an idea that he may have come up with himself but was not original, Constantijn continued. What was original, he wrote, was a white screen parallel to the wall that could be moved to and from and turned around. Apparently Drebbel used the camera to make himself

[33] Beeckman 1939–1953, 439–442 (appendix II). Jaeger denies this similarity but gives no further arguments. Jaeger 1922, 134 footnote 1.

[34] Borel introduced the claim that Sacharias Jansen invented the telescope and the microscope (Borel 1655). Within a few decades Christiaan Huygens had refuted the claim (see below); Zuidervaart 2010 shows how it persisted until the present day. Ruestow 1996, 7 repeats the claim that Drebbel obtained one from Jansen. Harting in his historical exposition also denied Drebbel's authorship: Harting 1850, 24–28.

[35] Beeckman 1939–1953, 440.

[36] Letter to IJsbrandt van Rietwijck in Alkmaar, around 1608–1609. Transcription Jaeger 1922, 110. 'UE. heeft mij voor desen geswegen 't verre sien gevonden bij den zoon van Mr. *Adriaen Thonissen*. Ick bidde laat mij weten wat daerin gedaen heeft. Ick hebbe oock vele excellente dingen daerin gevonden, soo ongelooflijke schijnen ende als tooverij geestimeert werden, waervan UE. hier een weinich wil gedencken.' The 'son' is often mistaken for Adriaan, including Keller 2008, 124; but in 390 footnote 397 Jacob is referred.

metamorphose by projecting colors and images on his robe.[37] The only thing that remained, he wrote, was that Drebbel would find a way to erect the images.[38] Huygens also brought a camera back home and would entertain visitors with the moving images on the screen.[39]

By this time, Drebbel had developed a whole family of visual instruments. Later he would be regarded as a model optics inventor alongside Galileo, Fontana and Rheita.[40] The Peiresc report describes all kinds of ways to cast images and light and view objects, like Archimedes-style mirrors to set fire at a distance and create floating images. Drebbel had developed a kind of spotlight: a configuration of multiple lenses with which candle light could be projected over a distance to make objects visible at night. The Peiresc report described an instrument 'that multiplied the light of a star in such a way that a letter can be read at night'.[41] Together with the camera and the microscope, instruments like these create the impression of a Drebbel dabbling with the set-ups and effects of his instruments. Combining and recombining the projection of images through a pinhole, the telescopic effects of the tube of long vision; reversing imagery and vision, inwards and outwards. Although the sources do not allow a full genealogy we can see Drebbel's microscope as the off-spring of the phantasmagoria camera and the telescope, and next of kin to the spotlight.

6.3 Quarters of the Magia Naturalis

With Drebbel too, terms like image, light and optical acquire a much broader meaning than the restricted dioptrical domain. Like Burggrav's lamp of light, Drebbel built all kinds of instruments that manipulated and utilized light. Peiresc described an artificial sun, 'that is to say a perpetual fire that always burns and lights.'[42] In Drebbel's ontology fire and light were equal, and fire was the quoin of the round of

[37] Jaeger 1922, 111 calls the instrument in the 1609 letter a magic lantern or camera lucida but this interpretation is very problematic for several reasons: the description is evidently camera obscura like and no 'slides' are projected on an outer screen; it is chronologically implausible.

[38] Worp 1897, 119. '…; levis operae instrumentum, quo quae foris obiiciuntur, sole valido illustrata, in cubiculum exquisite occlusum speciem sui intromittunt. Aperto tenui foramine maiores utebantur; primum Drebbelium applicasse fenestellam vitream orbicularem creditum fuit, sed et hanc superiorum industriae debet, nisi nescium veteris inventi dicas per se, quasi iam recentis, autorem extitisse, quod accidere non raro sciunt, qui, quo candore decet, omnium saeculorum laudes ex aequo librant. Illud constat, candidae tabellae cum pariete παραλληλισμόν, motum item et accessum et recessum et in omnem plagam facilem obversionem solius Drebbelii esse; perfectâ nunc iucundissimi longe atque utilissimi spectaculi machinâ, si, quas decussatis specierum radiis inversas imagines dat, erigat tandem Drebbelius meus et contractum diu nomen aliquando solvat.'

[39] Worp 1897, 83–84.

[40] Keller 2008, 39–40.

[41] Humbert 1955, 155. 'de multiplier la lumière d'une estoille en sorte qu'elle puisse faire lisre une lettre de nuict'

[42] Jaeger 1922, 132.

transformations of the four elements.[43] In his dynamics of atmospheric evaporations and condensations fire was the principal power of change; as an active agent rather than a substance. Drebbel recreated these processes in instruments like the famous perpetuum mobile and the self-regulating oven. Light was more directly involved in the clock that readjusted itself to the sun and the harpsichord that played to the light of the sun.[44] Vera Keller dubs the renowned perpetuum mobile 'cosmoscope': an instrument providing a look into the cosmos.[45] Huygens described it as a glass spiral that contained a fluid that imitated the tides in an ongoing periodical movement.[46] He compared it to the rarefaction and condensation through the change of temperature, not unlike the way we would nowadays say that Drebbel's living instruments utilized changes of atmospheric pressure.[47] In his oeuvre the cosmoscope and the microscope are close: heat and light being manifestations of a similar fiery quality inherent to matter and the instruments means to bring forth this latent fire.

In 1607 Gerrit Pietersz. Schagen (1573–1616) introduced his fellow townsman with these words: 'But this philosopher from Alkmaar can demonstrate [the principles of natural phenomena] not just with reason but also with living instruments.'[48] Drebbel's instruments channeled the powers of nature. Or rather: they emulated the powers of nature, for they were simulacra of natural phenomena, a microcosm displaying the workings of the world. In an early statement he wrote:

> thus also makes all kinds of instruments that play in its time and in all what can be made for a time by descending weight or by springs, by running waters, by wind, or by fire; that can be made by this knowledge for eternity.[49]

Drebbel's living instruments inhabit the world of *Magia Naturalis*. Like the designs of Della Porta the instruments of Drebbel were means to bring about effects. These effects were in fact phenomena replicated by human means and in a controlled manner. The creation of effects implied the understanding of phenomena thus recreated. In this phenomenological conception of understanding, instruments that imitated natural processes were the key to the workings of nature. Drebbel claimed to be able to explain the meteorological dynamics of the atmosphere by means of the thunder and lightning he artificially produced.[50]

---

[43] Drebbel 1621 (in particular the first three chapters; see also Keller's translation, Keller 2008, 508–524); Keller 2008, 44–46; 377–388.

[44] Drebbel described these as early as 1613: Jaeger 1922, 100–101; 110; 125–126.

[45] Keller 2010, 41–43.

[46] Worp 1897, 116–117.

[47] Borrelli 2008; Keller 2013.

[48] Drebbel 1607, n.p. laudatory epistle: 'Maer desen Alckmaersche Philosooph can 't selfde niet alleen met reden / maer oock met levendige Instrumenten bewijsen.' Compare the translation in Keller 2008, 499.

[49] Drebbel 1607, n.p.: 'also oock maeck allerley Instrumenten / die eeuwelijck spelen op haer tijdt/ en in summa wat voor een tijdt ghemaeckt kan werden / door dalent gewicht / of door springhveeren / door loopende wateren / door wint / oft door vier / dat kan ghemaeckt worden door dese kennis voor eeuwelijck'. Compare the translation in Keller 2008, 502.

[50] Keller 2008, 166

Arianna Borrelli has given a penetrating account of Della Porta's methods of inquiry. Employing a concept proposed by Bertoloni Meli she characterizes it as 'thinking with objects'.[51] Della Porta built upon the understanding of the effects of an artifact to acquire an understanding of a related, more complex one. In this way he worked in optics from plane to spherical mirrors, and from reflection to refraction, and eventually to lenses and their configurations. The crux here is that understanding is based on the properties and manipulation of images as a whole, rather than the analysis of their make-up by tracing the paths of rays. (And we need to bear in mind that these are images as perceived by the observer, and not the Keplerian pictures painted by rays.) This holistic and perceptual understanding of images can also be recognized in the accounts and reflections of Drebbel. The lineage of his optical instruments suggests that he extended his command over imagery from instrument to instrument: transforming the casting of camera images into streaming a bundle of light; reconfiguring a weatherglass into a light organ; and so on. The assumption that Drebbel was 'thinking with objects' in similar fashion as Della Porta is confirmed by his artefactual epistemology in which the understanding of natural dynamics consists of its instrumental emulation.

Still, Drebbel's writings are scarce. They do not reveal much about his instruments beyond pitches about their wonderful effects. Most of the information about the design and workings of his instruments comes from eyewitness accounts such as Peiresc and Huygens. In his own writings – basically a single exposition on the nature of elements – Drebbel focused on his natural philosophy.[52] They do not offer a very explicit account of light, images and perception. Light is largely subsumed under the category of fire, the central element in his conception of atmospheric dynamics. Drebbel and Della Porta stood in the same tradition of sixteenth-century criticism of Aristotelian theories of meteors.[53] Both set great store by a 'thermodynamic' explanation of winds substantiated in an experiment with an inverted glass vessel – showing barometric action in modern words. In this artifactual and metereological account of nature Drebbel focused on ontology in comparison to the precise epistemological steps of Della Porta.

Drebbel was connected to a circle of Low German alchemists with Hartmann as a prominent figure. The network also played a prominent role in the dissemination of his ideas and the continuation of his reputation.[54] Burggrav was an important promotor of Drebbel's philosophy, publishing German and Latin editions of his book. Even Andreas Libavius (ca. 1555–1616), who was quite critical of the group's opinions, was respectful of the artifacts and theories of Drebbel.[55] He did, however, reject Drebbel's conviction that his living machines provided a key to the cosmos.

[51] Bertoloni Meli, 2006, 1–17. Borrelli 2014, 41–46.

[52] For a detailed discussion of Drebbel's natural philosophy, as well as his ambitions and influence as natural philosopher, see Keller 2008.

[53] Borrelli 2008, 78–85.

[54] The Herborn scholar Alstedt adopted his ideas and the Kufflers became prominent members of the Hartlib circles in Northern Europe. Keller 2008, 429–461; Young, 2006.

[55] Keller 2008, 214; 406.

These northern protagonists introduced a decidedly vitalistic element in natural magic. Drebbel's cosmoscope channeled the perpetual, automotive movements of the tides. According to Burggrav it was driven by the 'magnetic spark of the Anima Mundi.[56] His own lamp of life functioned on the principle that the human blood attracts the vital celestial forces that direct individual health.[57] They also gave a spiritual twist to their inquisitive practices, fusing natural and divine knowledge in reading the Book of Nature alongside the Scripture. Drebbel strongly adhered to this religiously anchored epistemology. He acquired understanding of the workings of nature by means of devising ingenious instruments that disclosed the secrets of nature. This operational way of knowing was grounded in manual labor performed under the guidance of God. Adoration by instruments, so to say.

Characterizing Drebbel as a chymist is hazardous. He was trained as an engraver and originally he was active in hydraulic projects, building fountains and acquiring patents for pumps and chimneys around 1600.[58] He was generally referred to as an engineer and mathematician, and his instruments were called mathematical.[59] The baptism of Drebbel's fire regiment as 'thermometer' took place in the *Récréations Mathématiques*, a widespread collection of 'secrets' that went through various editions after 1624.[60] Calling metereological instruments mathematical is a fine example of shifting categories from our perspective. In general mathematics was commonly ranked among secret knowledge and natural magic, with Archimedes as the paragon of ingenuity.[61] In his reflection on the original wisdom – written around 1600 – Simon Stevin listed magic with arithmetic and chymistry among the instances of pre-classical knowledge.[62] As a hydraulic engineer Drebbel fits very well in the lineage of Dutch virtuosi like Stevin and Beeckman. Stevin may well provide a model of a vernacular philosopher for Drebbel, but historiography usually sets them apart.[63] Already in 1612 Drebbel was considered as a singular example of Dutch ingenuity.[64] Gerrit Schagen introduced him in 1607 as a nova, like the nova observed by Metius' father Anthonisz., the original Dutch engineer from Alkmaar.[65]

---

[56] Keller 2008, 391.

[57] Keller 2008, 111 footnote 271.

[58] Tierie 1932, 31–36; see also Jaeger 1921 and Keller 2008.

[59] Borrelli 2008, 111–113 cites a description of the experiment as a mathematical marvel and Drebbel as a smart mathematician.

[60] Keller 2013, 244–245.

[61] Schneider 1998, 1–26; Keller 2008, 214; 325.

[62] Stevin 1608, 9–16 (in 'Bepaling 6' of 'Vant Eertclootschrift' in 'Vant Weereltschrift').

[63] Keller 2008, 25 footnote 78. Stevin hardly figures in Keller's account, instead she presents Coornhert as a possible model.

[64] Keller 2008, 135.

[65] Drebbel 1607, n.p.: laudatory epistle by Gerrit van Schagen.

6.4 Porta Readings in the North

The case of Drebbel shows that there was definitely room for natural magic in the Low Countries.[66] His work was well-known and well-received, and throughout the seventeenth century his heritage is encountered. Likewise, Della Porta's works were well circulated and respected among northern readers.[67] As early as 1566 the *Magia Naturalis* was translated into Dutch and published by Plantijn in Antwerp. 'Magic, or the miraculous works of nature', contained the four books of the first edition of 1558. The second edition of 1589 was not translated, though. From the 1650s two editions of the Dutch *Magia* were published in Leiden but they were based on the 1566 translation, confined to the contents of the first four-volume edition of the *Magia*. Dutch readers had to rely on the original Latin for Della Porta's extension of his optical account to spheres and lenses of book XVII of the twenty-volume sequel. The original editions of Della Porta's publications were found in many libraries, though. Constantijn Huygens owned both the *Magia Naturalis* and *De Refractione*; Beeckman used the *Magia* as well as the *Pneumatica*. Although his work and reputation were known, Della Porta does not seem to have been a particular focus of interest. His name was mentioned among others. Someone like Huygens referred to Della Porta a few times in his correspondence but merely in the passing. The influence of the *Magia Naturalis* may have been more indirect. Through the general notion of natural magic as it appears in the *Récréations Mathematiques*; and along lines of specific topics like the weather glass.

Isaac Beeckman was very interested in natural magic. Like Drebbel he had engaged in hydraulic engineering projects, before embarking upon a scholarly career as a headmaster. From an early point on he was engaged with Drebbel's work; he read a copy the rare first edition of *Natuere*.[68] He was closely connected to the Drebbel circles and is – as we have seen – an important source of information on his instruments.[69] Beeckman also knew Della Porta's work well, having access to works like this in the library of the Dordrecht minister Colvius. On one occasion he discussed Della Porta's account of the phases of the Moon and their relationship to earthly humidity, praising the Neapolitan for his depth compared to others.[70] In

[66] In Franeker the chymistry of Burgrav and Hartmann seems to have faded somewhat in the background after the 1620s. Adriaan Metius appears to have lost part of his interest after the death of Sternsee (and the disappearance of a substantial part of his funds). This is at least the suggestion that his funeral orator Winsemius makes: alchemy had temporary fascination of Metius. Winsemius, by the way, resided in the Franeker castle but took no interest in the chymical laboratory of its previous owner. The further history of chymistry in the Low Countries need not occupy us here, but is well worth being told – including the vital uses of light.

[67] For an overview see Gemert, 2008.

[68] Beeckman 1939–1953, I, 346; II, 25; 44; 122. His father kept him up to date on developments and the book may also have been among their Middelburg circles: Berkel 2013, 30–31; 49–51. See also Keller 2008, 58. 1607.

[69] Beeckman 1939–1953, III, 302.

[70] Beeckman 1939–1953, II, 34–35.

optics too magical traits can be discerned in Beeckman's reflections, in particular in his accounts of colors. He regarded color as a clue to the nature of materials that are transformed under the influence of fiery reactions.[71] In his work on lenses and telescopes he took, however, a different direction than Drebbel and Della Porta. Beeckman is one of the few who engaged deeply in dioptrics in the Low Countries in the early seventeenth century. During the final years of his life he was intensively occupied with the grinding of lenses and the construction of telescopes. In his theoretical and experiential reflections he showed much originality and inventiveness, but he developed a rather geometrical approach along Keplerian lines.[72] As a result, in Beeckman's account of optical effects there was little room for a phenomenological approach like that of Drebbel and Della Porta.

The phenomenological approach to optics resonated much better with Constantijn Huygens. Upon his return to the Republic he brought a Drebbel camera and microscope with him to entertain and inspire his acquaintances and visitors. He urged painters to make use of the microscope to paint a new world and employ the camera as a way to bring painting to life.[73] He suggested scholars to study refraction and supported Descartes' efforts to build a flawless telescope.[74] During his long life Huygens became one of the most prominent cultural intermediaries of the Republic. His case illustrates the relative marginality of Della Porta in Dutch culture. He knew and esteemed the *Magia Naturalis*, but his unreserved praise went to the ingenuity of Drebbel. In general, it seems that drebbelian conceptions and circles were dominant in northern natural magic and overshadowed Della Porta.

Constantijn passed on his enthusiasm for optics this his sons Constantijn jr. (1628–1697) and Christiaan (1629–1695) (he wrote his memento on his youth in the weeks after the latter's birth). Together they would grind lenses, build telescopes, and observe the heavens.[75] Christiaan would of course acquire fame with his mathematical and philosophical writings on light and telescopes but he also continued another part of his father's optical interest: the drebbelian camera. In 1662 his father was at the Paris court involved in laborious negotiations for the Oranges. With a view to entertain his audience he asked his son to send a 'lantern with two or three pictures of which it makes the representation'.[76] This was a magic lantern and Chrstiaan had invented it somewhere in the 1650s.[77] Rather than projecting light on a sheet inside the camera, it projected light through a picture on glass upon a surface

[71] Dijksterhuis 2016.

[72] Dijksterhuis 2010.

[73] Huygens 1911, 94 (Letter 143, to his parents April 1622).

[74] Dijksterhuis 2011, 100–103; Dijksterhuis 2007, 65–67.

[75] Dijksterhuis 2004, 53–63; 214–216.

[76] Huygens 1888–1950, IV, 102 (letter 1001; to his brother Lodewijk, 5 April 1662). 'Voila encore une autre commission que mon Pere me donne, de luy ajuster une lanterne avec 2 ou 3 diverses peintures dont elle face la representation.'

[77] Huygens 1888–1950, XXII, 196–197; 521–523.

outside. The historical link between Christiaan's magic lantern and the Drebbelian camera of his father requires further investigation.[78]

Christiaan's attitude differed considerably from his father's, though. He reluctantly complied with the request and secretly instructed his brother Lodewijk to leave out a crucial lens in order to more or less sabotage it.[79] Apparently Christiaan did not want to be associated with an instrument that he considered nothing but a playful entertainment. Historiography has strictly separated the natural magic of Drebbel from the science of Christiaan Huygens. The perpetuum mobile is often invoked to illustrate the chasm. However, even in the case of Christiaan Huygens things are not as simple as they appear. He may have rejected the possibility of a perpetuum mobile qua mechanics but he did not exclude the possibility as regards heat or magnetism.[80] That is: machines could very well run on their own when employing the hidden forces of nature. Natural magic was as real for Christiaan as it was for his father.

Huygens had been reading Della Porta quite well. Although he never discussed the ideas of *Magia Naturalis* and *De Refractione* extensively, he approvingly mentioned them at several places.[81] For example, he named Della Porta as one of the pioneers in the 'measure of refraction' – that is the mathematical regularity of refraction angles – with Maurolyco and Kepler. Furthermore, in the theory of vision – that Huygens never really elaborated but did make some notes about in his manuscripts – Della Porta figures prominently with his account of images appearing in the dark. Huygens also noted that Della Porta has the image in the crystalline lens as contrasted to the retinal image of Kepler.

Around 1692, at the end of his life and not that long after the death of his father in 1685, Christiaan was reading the *Dioptrica Nova* of William Molyneux, making all kinds of comments and corrections.[82] At one point he noted that Molyneux, on the basis of Borel and Sirturi promoted Sacharias Janssen and Fontana as inventors of the telescope. Huygens had refuted that claim himself: some time earlier he had done some archival work and found out that the authorship of Janssen was

---

[78] The phantasmagorical projecting of colors and objects on a screen by Drebbel is interesting. It suggests a continuity from Drebbel to Christiaan Huygens, Walgenstein and Kircher. The editors of the *Oeuvres Complètes de Christiaan Huygens* refer to Constantijn's interest but do not discuss it further: Huygens 1888–1950, XXII, 521–523. Wagenaar's substantive history of the camera from 1979 does not mention it; Hankins 1995 (chapter 2: 'The Magic Lantern and the Art of Demonstration') 43–48 mentions it but does not pursue the suggestion. Neither does Steadman, who apparently has not inspected the details of Constantijn's description; his suggestion that the camera depicted by Kircher is akin to Drebbel's is significant. Steadman 2001, 17–19. Keller merely mentions the camera, without considering the details of the set-up. Keller 2008, 233 footnote 540; 471 calling it magic lantern; on p. 471 explicitly. Keller 2008, 21 footnote 64 raises the question whether it was an arrangement of mirrors rather than a lantern, but does not elaborate on it.

[79] Huygens 1888–1950, IV, 109–110 (letter 1004; to his brother Lodewijk, 12 April 1662).

[80] Huygens 1888–1950, XIX, 553; 575–581; 591–603.

[81] Huygens 1888–1950, XIII, 437; 588–590; 740–750.

[82] Huygens 1888–1950, XIII, 826–844.

manufactured.[83] Huygens elaborated a brief history of the telescope in the 1680s when he was preparing a treatise on dioptrics. Besides the theory of the telescope – that he had elaborated in the 1650s but never published – it also included the microscope. He named Drebbel as the true inventor of that instrument. The true inventors of the telescope were Lipperhey and 'the brother of Metius' he concluded. But Della Porta deserved the credit for having been the first to suggest the idea of the telescope. If he had not made sufficient progress, this was because he only knew the rudiments of the art of optics as Huygens knew it.[84] This appreciation for Della Porta went back to his first plans for a treatise on dioptrics in 1672: 'Porta was the first who had begun.'[85]

References

Beeckman, Isaac. 1939–1953. *Journal tenu par Isaac Beeckman de 1604 à 1634*, ed. C. de Waard. The Hague: Martinus Nijhoff.

Berkel, Klaas van. 2013. *Isaac Beeckman on matter and motion: Mechanical philosophy in the making*. Baltimore: The Johns Hopkins University Press.

Bertoloni Meli, Domenico. 2006. *Thinking with objects: The transformation of mechanics in the seventeenth century.* Baltimore: The Johns Hopkins Press.

Borel, Pierre. 1655. *De vero telescopii inventore. cum brevi omnium conspiciliorum historia; ubi de eorum confectione, ac usu, seu de effectibus agitur, novaque quaedam circa ea proponuntur, accessit etiam centuria observationum microcospicarum*. The Hague: Adriaan Vlacq.

Borrelli, Arianna. 2008. The weatherglass and its observers in the early seventeenth century. In *Philosophies of technology: Francis Bacon and its contemporaries* (Intersections 11/1), ed. Claus Zittel, Gisela Engel, Nicole C. Karafyllis and Romano Nanni, 67–130. Leiden: Brill

———. 2014. Thinking with optical objects: glass spheres, lenses and refraction in Giovan Battista Della Porta's optical writings. *Journal of Early Modern Studies* 3: 39–61.

Burggrav, Johann Ernst. 1610. *Lampas Vitae et Mortis omniumque graviorum in Homine pathoon*. Leiden: Hendrik Haestens.

———. 1611. *Biolychnivm seu Lvcerna : Cum vita ejus, cui accensa est Mysticeì, vivens jugiter, cum morte ejusdem expirans, omnesq[ue] affectus graviores prodens*. Franeker: Uldricus Balck.

Dijksterhuis, Fokko Jan. 2004. *Lenses and waves. Christiaan Huygens and the mathematical science of optics in the seventeenth century*. Dordrecht: Springer.

———. 2007. Constructive thinking. A case for dioptrics. In *The mindful hand: Inquiry and invention from the late Renaissance to early industrialisation*, ed. Lissa Roberts et al., 59–82. Amsterdam: Edita.

———. 2010. Labour on lenses. Isaac Beeckman's notes on lens making. In *The origins of the telescope*, ed. Albert van Helden, Sven Dupré, Rob van Gent, and Huib Zuidervaart, 257–270. Amsterdam: KNAW Press.

[83] Huygens 1888–1950, XIII, 436–437; 591–593.

[84] Huygens 1888–1950, XIII, 436–437.

[85] Huygens 1888–1950, XIII, 738–741. 'B. Porta est le premier qui a commencè'. Huygens added that reading of Della Porta may have assisted the Dutch achievement: 'Progres en Hollande peut estre par la lecture de Porta.' On the dating of these manuscript notes see Dijksterhuis 2004, 92 and 140.

———. 2011. Moving around the Ellipse. Conic sections in Leiden, 1620–1660. In *Silent messengers: The circulation of material objects of knowledge in the early modern low countries*, ed. Sven Dupré and Christoph Lüthy, 89–124. Berlin: Lit.
Dijksterhuis, Fokko Jan. 2016. Varieties of theories in the color worlds of the early seventeenth century. In *Early modern color worlds*, ed. Tawrin Baker, Sven Dupré, Sachiko Kusukawa, and Karin Leonhard, 227–247. Leiden: Brill.
Dijkstra, Arjen. 2012. *Between Academics and Idiots. A Cultural History of Mathematics In The Dutch Province Of Friesland (1600–1700)*. Dissertation University of Twente
Drebbel, Cornelis. 1607. *Wonder-Vondt van de Eeuwighe Beweghingh*. Alkmaar: Jacob de Meester.
———. 1621. *Een Kort Tractaet van de Natuere Der Elementen*. Haarlem: Vincent Casteleyn.
Dupré, Sven. 2005. Ausonio's mirrors and Galileo's lenses: The telescope and sixteenth-century practical optical knowledge. *Galilaeana: Journal of Galilean Studies* 2: 145–180.
———. 2006. Visualization in Renaissance optics: The function of geometrical diagrams and pictures in the transmission of practical knowledge. In *Transmitting knowledge: Words, images and instruments in early modern europe*, ed. Sachiko Kusukawa and Ian Maclean, 11–39. Oxford: Oxford University Press.
Gemert, Guillaume van. 2008. Giovan Battista Della Porta in den Niederlanden. Ansätze zu einer Rezeptionsgeschichte. *Morgen-Glantz* 18: 29–61.
Hankins, Thomas, and Robert Silverman. 1995. *Instruments and the imagination*. Princeton: Princeton University Press.
Harting, Pieter. 1850. *Het Mikroskoop, deszelfs gebruik, geschiedenis en tegenwoordige toestand*. Utrecht: Van Paddenburg & Comp.
Humbert, Pierre. 1951. Peiresc et le Microscope. *Revue d'histoire des sciences et de leurs applications* 4–2: 154–158.
Huygens, Christiaan. 1888–1950. *Oeuvres complètes de Christiaan Huygens*. Den Haag: Martinus Nijhoff.
Huygens, Constantijn. 1911. In *Briefwisseling. Deel 1: 1608–1634*, ed. J.A. Worp. Den Haag: Martinus Nijhoff.
———. 1987. *Mijn jeugd* (trans., ed. C.L. Heesakkers). Amsterdam: Querido.
Jaeger, F.M. 1922. *Cornelis Drebbel en zijne tijdgenooten*. Groningen: Noordhoff.
Jorink, Eric. 2010. *Reading the book of nature in the Dutch golden age, 1575–1715*. Leiden: Brill.
Keller, Vera. 2008. *Cornelis Drebbel (1572–1633): Fame and the making of modernity*. Dissertation Princeton University.
———. 2010. Drebbel's living instruments, Hartmann's Microcosm, and Libavius's Thelesmos: Epistemic machines before Descartes. *History of Science* 48-1: 39–74.
———. 2013. Re-entangling the Thermometer: Cornelis Drebbel's description of his self- regulating oven, the regiment of fire, and the early history of temperature. *Nuncius* 28: 243–275.
McKeon, Robert M. 1986. Le renouvellement de l'astronomie de précision de Tycho Brahe a Jean Picard. In *Picard et les debuts de l'astronomie de précision au XVIIe siècle*, ed. Guy Picolet, 119–131. Paris: CNRS.
Reeves, Eileen. 2008. *Galileo's glassworks. The telescope and the mirror*. Cambridge: Harvard University Press.
Ronchi, Vasco. 1954. Du 'De Refractione' au 'De Telescopio' de G.B. Della Porta. *Revue d'Histoire des Sciences* 7: 34–59.
Ruestow, Edward. 1996. *The microscope and the Dutch Republic. The shaping of discovery*. Cambridge: Cambridge University Press.
Salloch, Sabine. 2006. *Das hessische Medizinalwesen unter den Landgrafen Wilhelm IV. und Moritz dem Gelehrten. Rolle und Wirken der fürstlichen Leibärzte*. Dissertation Philipps-Universität Marburg.
Schneider, Ivo. 1998. *Johannes Faulhaber 1580–1635. Rechenmeister in einer Welt des Umbruchs*. Basel: Birkhäuser.
Steadman, Philip. 2001. *Vermeer's camera: Uncovering the truth behind the masterpieces*. Oxford: Oxford University Press.

Stevin, Simon. 1608. *Wisconstige Gedachtenissen,* vol. 1. Leiden: Jan Bouwensz.

Tierie, Gerrit. 1932. *Cornelis Drebbel (1572–1633).* Paris: Amsterdam.

Wagenaar, Willem. 1979. The true inventor of the magic lantern: Kircher, Walgenstein or Huygens? *Janus: archives internationales pour l'histoire de la médecine et pour la géographie médicale* 66: 193–207.

Winkler, Mary G., and Albert Van Helden. 1993. Johannes Hevelius and the visual language of astronomy. In *Renaissance and revolution: Humanists, scholars, craftsmen and natural philosophers in early modern Europe*, ed. J.V. Field and Frank A.J.L. James, 97–116. Cambridge: Cambridge University Press.

Worp, Jacob Adolf. 1897. Fragment eener autobiographie van Constantijn Huygens. *Bijdragen en Mededeelingen van het Historisch Genootschap* 18: 1–122.

Young, John T. 2006. *Faith, medical alchemy, and natural philosophy: Johann Moriaen, Reformed intelligencer, and the Hartlib circle.* Aldershot: Ashgate.

Zuidervaart, Huib. 2010. The 'true inventor' of the telescope. A survey of 400 years of debate. In *The origins of the telescope*, ed. Albert van Helden, Sven Dupré, Rob van Gent, and Huib Zuidervaart, 9–44. Amsterdam: KNAW Press.

Chapter 7
Using Invariances in Geometrical Diagrams: Della Porta, Kepler and Descartes on Refraction

Albrecht Heeffer

Abstract In this paper, I will demonstrate how geometrical diagrams on refraction were instrumental in the discovery of the sine law of refraction. In particular, I will show how a specific diagram in the *Paralipomena* assisted Kepler in looking for invariances of proportions under different angles of incidence. Eventually, Kepler failed in finding a quantitative law of refraction, but it will be shown that his basic hypothesis and methodology can lead to the discovery of a quantitative law and that probably this was Descartes' path to the discovery of the sine law. Both Kepler and Descartes could build on a tradition of geometrical reasoning which accounted for co-exact properties in geometrical diagrams. Della Porta was the first to recognize such properties in diagrams dealing with refraction.

7.1 Reasoning with Geometrical Diagrams

The precise epistemological functions of diagrams have come under increasing scrutiny in recent writings within the philosophy of mathematical practice. Such a late interest is somewhat surprising since diagrams were already indispensable to Euclidean geometry. In the past, a mathematical diagram was at best seen as a tool with some heuristic value. Diagrammatic reasoning is now considered a crucial part of mathematical practice.[1] Current research attributes a much wider role to mathematical diagrams, considering them cognitive tools which allow to delegate some of the cognitive load to the external environment. Mathematicians perform "epistemic actions" on diagrams, which are defined in the book by Kirsh and Maglio,[2] as "actions that are performed to uncover information that is hidden or hard to compute mentally". While diagrams obviously resort to our visual capabilities, they are not limited to mere visual actions but also imply kinetic mental operations. This

[1] For a recent survey on diagrams in the philosophy of mathematical practice see Giardino (2017).
[2] Kirsh and Maglio (1994).

A. Heeffer (✉)
Ghent University, Ghent, Belgium
e-mail: Albrecht.Heeffer@UGent.be

A. Borrelli et al. (eds.), *The Optics of Giambattista Della Porta (ca. 1535–1615): A Reassessment*, Archimedes 44, DOI 10.1007/978-3-319-50215-1_7

kinetic aspect is one of the two features of reasoning with geometrical diagrams that will become important in our approach to diagrams on refraction.

A second aspect of geometrical diagrams that will feature in our discussion on refraction is their ambiguity. Diagrams have to be interpreted, as they can be over-determined or underdetermined for their purpose. Take for example an abstract Euclidean diagram as discussed by John Mumma (see Fig. 7.1).[3] The diagram shows a circle *D* in which lies a point *C* so that line *CE* intersects the circumference of the circle. We also see the radius *AB* connecting the center of the circle with point *B* on its circumference. Some of these elements, such as center point *A* are determined. However, point *B* can be any point on the circle and segment *CE* can cut the circle at any other point. So, depending on the specific reasoning involved, the positions of *C*, *E* and *B* are indeterminate.

Kenneth Manders, in a philosophical study on the semantics of Euclidean diagrams, talks about 'fallacies of diagram use', referring to the misunderstanding that exact conditions can be just read off from the diagram.[4] To assess what lettered diagrams determine, he distinguishes *exact* from *co-exact* attributes. Co-exact attributes are "those conditions which are unaffected by some range of every continuous variation of a specified diagram". In the example of Fig. 7.1, the exact attributes determine that *A* is the center point and that *C* is located within the circle. All others are co-exact attributes.

These apparent ambiguities and the over- or underdetermination of properties of geometrical diagrams are productive in the way they function as cognitive tools. As an example, let us look at a typical diagram used in the proof that the sum of the angles of every triangle equals 180° (see Fig. 7.2). The proof depends on the construction of line *DE* parallel to the base of triangle *ABC*, touching point *A*. Because of Euclid, I prop. 29, angles *DAB* and *EAC* equal the alternate angles α and β. As can be read from the diagram, the sum of the angles α, β and γ equals 180°.

While this diagram shows one specific triangle – and as such – is overdetermined, it can be easily understood that this property of triangles remains unchanged

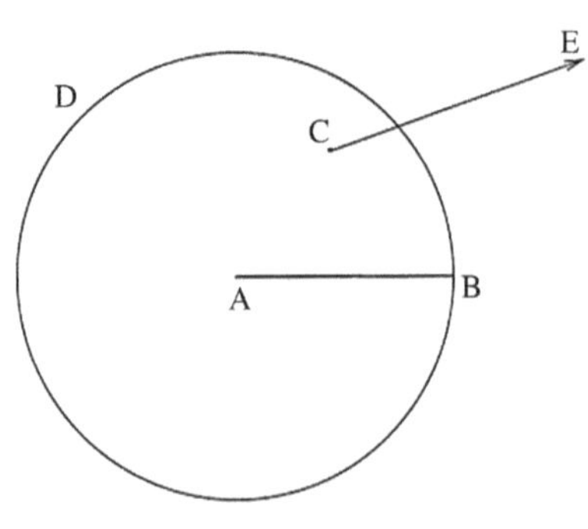

Fig. 7.1 An abstract Euclidean diagram with exact and co-exact properties

[3] Mumma (2010).

[4] Manders (2008).

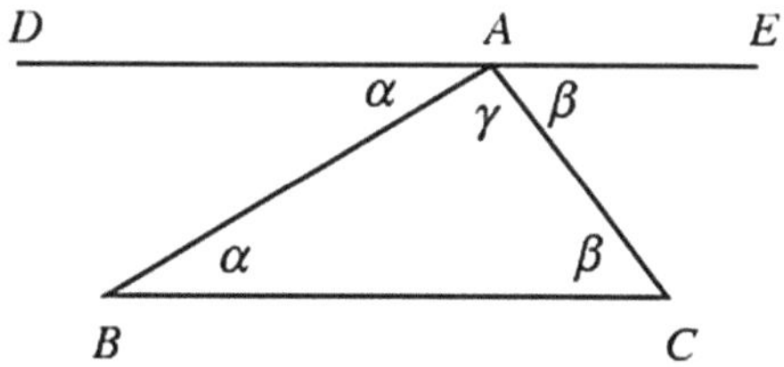

Fig. 7.2 The proof that the sum of the angles of a triangle equals 180°

for other angles. If angle α becomes larger, the shape of the triangle will change and angles β and/or γ will become smaller. However, the sum of the three angles α, β and γ remains invariant, which is an exact attribute of triangles. Our understanding of the proof thus depends on the invariance of co-exact attributes, while we mentally rotate the sides, change the angles and deform the shape of the triangle. Mental operations such as rotation, lateral movement, transformations, projections and resizing help us to discover invariances in diagrams. This comprehension of geometrical diagrams is essential to the Euclidean tradition and is shared with the Medieval Perspectivists as well as with Early-modern geometers.

7.2 Exact Attributes in Optical Diagrams

Optical diagrams differ in one important way from abstract geometrical diagrams in that they are constrained by physical laws. The angle of reflection equals the angle of incidence. This is a physical law from catoptrics which has to be respected in optical diagrams. Diagrams in dioptrics are even more constrained. Let us look at an archetypical diagram from Ptolemy's *Optics* V, 7–11, which depicts a diopter, an instrument to measure the angle of refraction for different media (shown in Fig. 7.3).[5]

Angle *AEZ* is the angle of incidence and is not determined. It can be any angle between zero and ninety degrees. However, the corresponding angle of refraction, angle *HEG*, is co-exact in relation to the angle of incidence. The co-exact property is determined by the densities of the media outside and inside of the instrument. Every diagram related to refraction is constrained in this way and should be drawn or at least understood in this way. For example, let us look at a diagram of light entering a glass globe as in Della Porta's *De refractione*, book II, prop. 11 (shown in Fig. 7.4). Point *A* is the eye and point *C* represents the object. Therefore, points *A* and *C* can be arbitrarily chosen. The center point *D* is an exact attribute. Hence, the cathetus *FDC* and points *E*, *B* are determined by the laws of refraction and are therefore co-exact attributes of the diagram.

With this short survey on recent scholarship on the semantics and cognitive functions of Euclidean diagrams we have provided the necessary background for our further analysis of optical diagrams. We would now like to propose a thesis for

[5] Smith (1996).

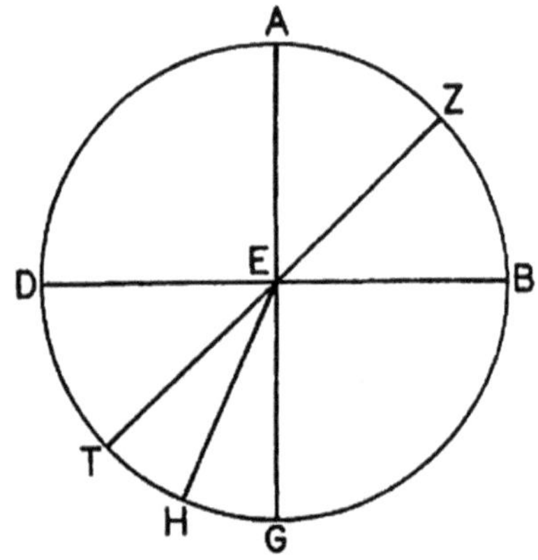

Fig. 7.3 Ptolemy's depiction of an instrument to measure refraction

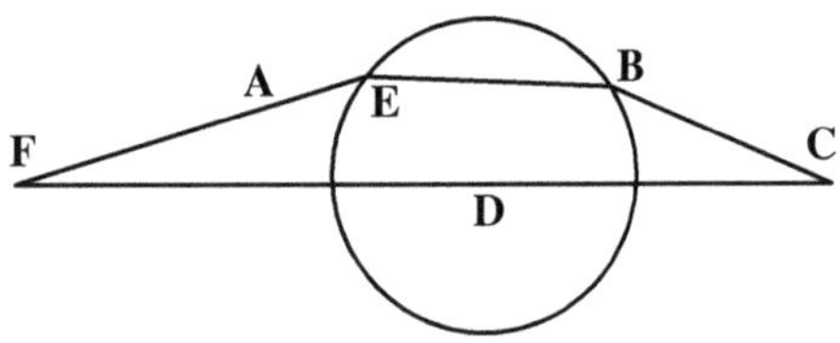

Fig. 7.4 Della Porta's figure in *De refractione*, book II, prop. 11

further discussion: A geometrical diagram of the measurement of refraction, when properly accounting for the exact and co-exact attributes, becomes itself an instrument that can lead to the discovery of the sine law of refraction.

The narrative of our story can briefly be summarized as follows: the first geometrical diagrams portraying refraction date back to Ptolemy. Witelo appears to have been the first who depicted variations in the angles of refraction by drawing several lines with different angles of inclination. The first attempt to account for exact attributes in diagrams representing refraction materializes in Della Porta's *De Refractione*. In his *Paralimonea*, Kepler is the first who consciously looks for invariances in diagrams which show different angles of inclination and refraction. And finally, Descartes arrives at the sine law of refraction by following Kepler's analysis, though he prefers not to disclose this path of discovery. We will now look into the historical development in detail, but we will do so in an anti-chronological way, starting with Descartes.

7.3 Descartes's Path to Discovery of the Sine Law

The literature on Descartes' discovery of the sine law is profuse. As soon as some years after Descartes' death, the Dutch scholar Isaac Vossius recognized in Descartes' explanation of the sine law an attempt to cover up his plagiarism from Snell. As a result, in Anglo-Saxon countries the sine law of refraction is named after

Snell rather than Descartes. Even Ernst Mach wrote "After actually reading these discussions in Chapter II of Descartes' *Dioptrices,* it will scarcely be assumed, even apart from the statements of Huygens, that Descartes discovered the law of refraction".[6] Much of these fierce reactions against Descartes can be explained by the particular way he presented the sine law in his *Discourse.*[7]

For the second chapter of his *Optics*, where the sine law is presented, he had the choice between two options. Either he could deduce the sine law from primary truths as an application of his method explained in his *Discourse*. Or he could proceed in a different way, more suited for explaining physical phenomena: starting from hypotheses that are founded on empirical grounds, and then deducing the causes. He favored the second option for reasons he explained in the *Discourse:*

> I have called them hypotheses only in order for you to know that I think I can deduce them from those first truths which I have explained above, but that I expressly wished not to do so, because of certain minds who imagine that they can comprehend in a day what has taken another person twenty years to think out, and I wish to prevent them from taking the opportunity to build some extravagant philosophy upon what they will believe to be my principles, and then attributing the blame to me.[8]

In fact, Descartes failed to find an explanation derived from first principles. Ironically, Fermat in an attempt to disprove Descartes in 1662, and without his intend, proceeded in the Cartesian way advocated in the *Discourse:* he deduced the sine law from the first principle of economy of nature. Fermat borrowed this principle from a book by Marin Cureau de la Chambre, called *Light.* In this book Cureau applied the principle of the shortest path to derive the law of reflection. Of course, naively applying this principle to refraction would imply that light rays continue in a straight line through more dense media. Fermat went further to define a principle of least action, accounting for the difference in resistance to light in different media. Light proceeds along the path of the shortest time. Although Descartes would have disagreed with this principle as it contradicted most of his theory in *le Monde,* he should have admired Fermat's dexterous deduction of the laws of nature from general principles using the method of maxima and minima.

In any case, not a single scholar on the history of optics believes that Descartes' explanation of the sine law follows his path of discovery. Neither has Descartes claimed that he did, while for his analysis of the rainbow in the part on meteorology, he did explicitly state that his explanation did follow his path to discovery.[9] So how then did Descartes arrive at the sine law of refraction?

---

[6] Mach (1926: 33).

[7] Heeffer (2006).

[8] AT (VI, 76, p. 61).

[9] From a letter from Descartes to Vatier, February 1638 (AT, I, 559–660): "Nor could I show the use of that method in the three treatises that I included, since it prescribes an order of investigation which is different enough from the order I believed I must use in order to explain them. However, I have given a sample of it in describing the rainbow, and if you take the trouble to reread it, I hope it will satisfy you better than it did the first time". Translation from Ariew (2000, 86).

In a classic article dating back to the nineteenth century, Paul Kramer conjectured that Descartes arrived at the sine law by purely geometrical means, in particular the problem of the anaclastic curve.[10] This problem can be formulated in several ways but the most suitable in the context of dioptrics is the question, what should be the shape of the surface of a glass object that refracts all parallel rays of light into a single point. The thesis that this problem leads to Descartes' discovery is well supported by contextual evidence within the Mersenne circle and has been adopted by several scholars on seventeenth century optics.[11] Sabra is probably the only scholar to follow Descartes' physical dynamical model of light and believes that this explanation, though based on some rather awkward analogies, was the path of discovery.[12] This second hypothesis is probably the least acceptable as it depends on the notion of the speed of light. Although Sabra recognizes that Descartes never used the word 'speed' in connection with light, his complete analysis of Descartes' proof is based on the assumption incorrectly attributed to Descartes "that the velocity of light is characteristic of the medium it is traversing".[13] A third hypothesis, which supposes that experiments with a refractometer led to the discovery of the sine law, has been proposed by William Shea.[14] Shea points out the relevance of Kepler's discussion of such device for Descartes in the *Dioptrice* of 1611. A. Mark Smith, in an extensive study sees in Descartes' theory of light and refraction a logical progression from Perspectivist principles: "Descartes's theory emerges as a natural continuation of the Perspectivist effort to narrow the gap between what remained an essential static *explanation* of light and a kinetic *description* of its physical activity".[15] He believes the cosine law emerges as a mathematical consequence of such approach which was then later reformulated in the sine form. A fifth and final hypothesis is by John Schuster, originally published in his PhD[16] and recently reformulated.[17] He argues that experiments with the measurement of refraction together with Claude Mydorge has lead Descartes to the formulation of the cosecant version of the law (see Fig. 7.5), which was hence reformulated to the sine version in relation to the anaclastic. Schuster's account is based on Descartes' physico-mathematical agenda. He describes how Descartes already studied Kepler's *Paralipomena* in 1620 (including the diagram shown in Fig. 7.8) but at that time "embraced an assumption which would have hindered his deducing a sine law of refraction. He held that in two media the normal components of the force of light are in a constant ratio".[18] Schuster argues, basing himself largely on the Mydorge 'letter', that first Descartes discovered the law, in what he terms the cosecant form, by

[10] Kramer (1882).

[11] Milhaud (1921), Buchdahl (1969), Malet (1990).

[12] Sabra (1967).

[13] Sabra (1967: 113).

[14] Shea (1991).

[15] Smith (1996: 46).

[16] Schuster (1978).

[17] Schuster (2013).

[18] Schuster (2013: 160).

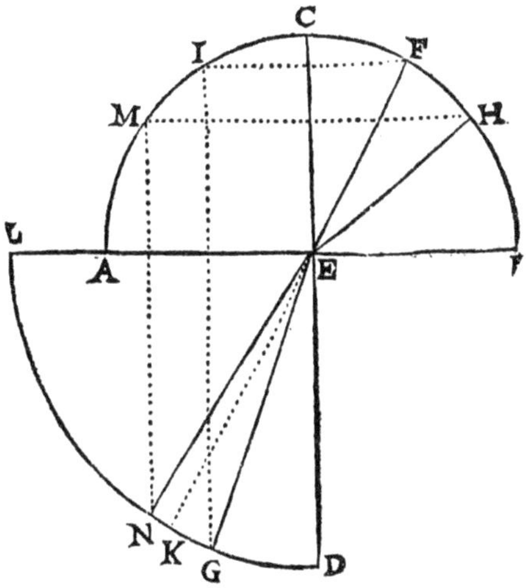

Fig. 7.5 Figure from a short treatise on refraction by Mydorge, copied by Mersenne. This is the actual figure drawn by Mydorge and pasted into manuscript NAF 5176 (used with permission from the Bibliothèque nationale de France)

purely geometrical optical means, based on the traditional image location principle and available data. Descartes then proceeded in what Schuster coins his 'physico-mathematical manner' and read out of the key diagram hints about the underlying dynamics of light, later somewhat unclearly expressed in his 'explanation' of the law of refraction in the *Dioptrics*. In a well documented appendix, Schuster offers a detailed genealogy of Descartes' lens theory based on the Mydorge 'letter'.

Remark that, apart from Sabra, all these authors agree on one main observation: Descartes discovered the sine law by geometrical deduction rather than by reasoning from physical causes and principles. Also, if we look at the argumentation of these authors in detail, geometrical diagrams play a crucial role in the way Descartes was supposed to arrive at his discovery, although this is not always stated explicitly. Both Shea and Schuster refer to experiments with the measurement of refraction, but at some point, the distinction between explanations using diagrams of the measurement of refraction and diagrams used for geometrical reasoning becomes obscure. As I will argue below, the diagram of an experiment with an instrument for measuring refraction becomes an instrument in itself.

I here propose the thesis that Descartes hit upon the sine law of refraction while studying the *Paralipomena*, in particular a diagram which is used by Kepler to look for invariances in geometrical proportions.[19] This thesis is in agreement with the general idea that geometrical reasoning formed the basis of the discovery and it also follows the importance attributed by Shea and Schuster to the measurement of refraction.

[19] Kepler (1604).

7.4 Descartes Confirming the Sine Law

Before moving to Kepler, we should look at historical records which allow us to set Descartes' discovery in its proper context. He published his explanation of the sine law in his *Discours de la méthode* in Leyden in 1637. It is generally acknowledged that he discovered the law much earlier. Less well known is that Descartes was not the first to publish the sine law. Mersenne had already published a version of the sine law in his 1636 edition of *Harmonie Universelle*, attributing the discovery to "l'un des plus excellens esprits de ce siècle", avoiding naming Descartes (see Fig. 7.6).[20] Also, Hérigone treated the sine law in the fifth part of his *Cursus mathematicus* which also appeared in 1637 but preceding Descartes' *Discourse*.[21]

In correspondence and manuscripts we find earlier references to Descartes' discovery. The earliest is an entry in Beeckman's *Journaal* dated October 1628 quoting "angulus refractionis Des Cartes exploratus".[22] The accompanying figure (see Fig. 7.7) is very similar to that published by Mersenne.

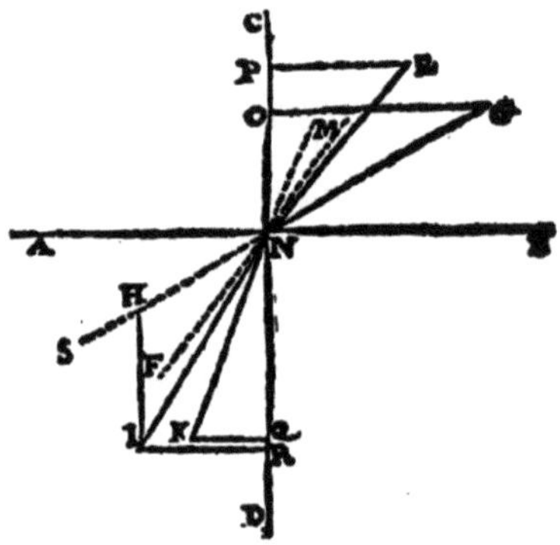

Fig. 7.6 Mersenne published the sine law in *Harmonie Universelle* (Mersenne 1636: 65)

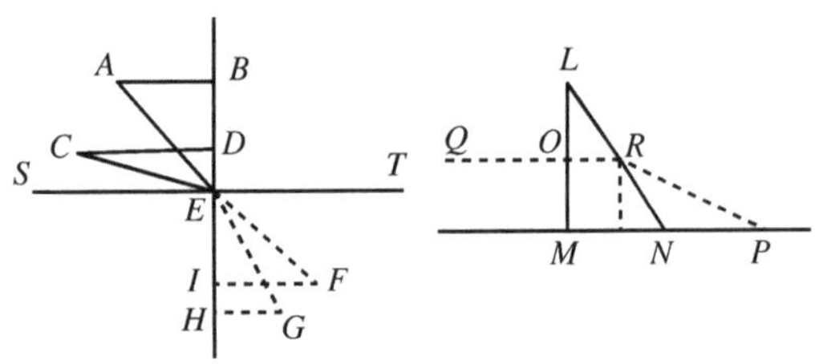

Fig. 7.7 The sine law in Beeckman's *Journaal* (De Waard 1939, III: 97)

[20] Mersenne (1636).

[21] Hérigone (1634–1644).

[22] De Waard (1939, III: 97).

In the correspondence of Descartes we find the law implicitly present in a communication to Ferrier of 8 October 1629 and a first explicit formulation in a letter to Mersenne of June 1632.

There is also a famous 'letter' of Mydorge to Mersenne expounding the law of refraction, dated by De Waard around February–March 1626.[23] However, closer inspection reveals that this is not a letter but a more formal representation of ideas which may have been discussed and communicated between Mersenne and Mydorge personally. The document is an autograph by Mersenne, copied from a treatise of Mydorge, possibly a draft of what should become book 8 of his *Prodromi catoptricorum et dioptricorum*, which was never published, or it could have been the treatise on refraction to which we have some references. This explains why the 'letter' remains undated and is not addressed to a person. This context should place this document in a rather different perspective. Schuster considers the 'letter' as a report by Mydorge sent to Mersenne on his joint experiments with Descartes and bases his hypothesis on the development of Descartes' lens theory on it.[24] However, nowhere in the document is there any hint to Descartes or shared experiences with Descartes. Instead, the short treatise on refraction builds further on Mydorge's own work on conic sections, particularly on book 1 and 2 of his *Prodromi* which was later published in 1631.[25] A letter by Ferrier to Descartes of 26 October 1629 reveals that Mydorge told Ferrier that "he alone discovered this". The small treatise on refraction also refers to Mersenne's *Quaestiones celeberrimae in genesim* of 1623, as noted by De Waard (CM, I, 414). The arguments used by Schuster, that Mydorge also cooperated with Descartes and that he credited Descartes with the discovery of the sine law does not provide sufficient evidence to the interpretation that this 'letter' is a report on their joint activities. In our perspective, the 'letter' should be interpreted more as a formal representation of previous discussions and common experiences by Mydorge and Mersenne.[26]

Mydorge's exact involvement in Descartes' discovery of refraction cannot be fully established but if not having discovered the law of refraction himself, he was aware of the sine law at a very early stage. In his short examination of problem 73

[23] CM (I, 404).

[24] Schuster (2013: 187–8).

[25] Mydorge (1631), more particularly book I, prop. 49, 51, concerning proposition II, and "now all this is done, it will be easy to describe the hyperbole," in proposition III, as he explained in book II, prop. 20, 26. Schuster does not mention the book in this context.

[26] The text by Mydorge is contained in a rough draft by Mersenne of an unpublished treatise on optics, kept as manuscript Fr. 5176 at the BNF in Paris. The manuscript is 30 pages long and is an unorganized and unnumbered collection of notes, bearing no resemblance to a published work by Mersenne. Lenoble (1957, 239) considers it a sketch to an abandoned continuation of the treatise on sound in the *Quaestiones in Genesim* (BNF. Lat. 17,261 and 17,262) although some of its propositions also appear in Mersenne's treatise on *Optics and Catoptrics* included in the second edition of Niceron's book on *Perspective* published in 1651.

of the popular *Récréations Mathématiques*, dealing with spectacles of pleasure he writes the following[27]:

> This noble subject of refraction whose nature has not been known, nor by the ancients, nor by modern philosophers or mathematicians up to the present, must now honor its discovery to a brave gentleman, one of our friends, most admirable in his knowledge and subtlety of mind, heretofore accomplished in all kinds of virtues, of which we hope he will provide us with his own account, among others in a treaty that he promised to make public (furthermore, as a consequence of which we engage ourselves *and our particular inventions*, [to find] ways of easily and reliably placing the theory into practice) which prevents us from saying any more here, which elsewhere affects what is commonly known as the glasses of Galileo, which is not better known than the other exact science, and can probably be better approached by trial and error (translation and emphasis mine).

The phrase 'our particular inventions' ('de nos particulieres inventions') may refer to the discovery of the sine law, but "reliably placing the theory into practice" rather suggests the application of the sine law to some geometrical findings, such as the solution to the anaclastic. This quote has been used by Schuster as an argument that Mydorge credits Descartes for the discovery of the sine law.[28] Now, interestingly this quote does not appear in the rare 1626 edition. We can therefore assume that when Mydorge was studying the *Récréations Mathématiques* at the end of 1625 until the beginning of 1626, Descartes had not yet discovered the sine law, or at least that Mydorge did not yet know about Descartes' discovery. So, Schuster's dating of the treatise by Mydorge discussed earlier[29] as "very plausibly as early as 1626/1627—but not earlier"[30] is hereby confirmed as the *terminus a quo*. The *terminus ante quem* is Oct 1628 but could be earlier depending on the date of Mydorge's small treatise on refraction. Costabel dates the document after the appearance of Mydorge's 1631 publication, but his and some other later datings stem from the observation that Mydorge included references to his own *Catoptrics*. However, we know that he had drafted and probably already completed the first two books several years before its publication. Mersenne refers to it already in his *Quaestiones celeberrimae in Genesim* of 1623, as does Mydorge in his revision of notes on the *Récréations Mathématiques*, on which he worked in 1628. The publication of his *Examen* was postponed[31] because an expanded edition of the *Récréations Mathématiques* appeared in 1628 by Charles Osmont in Rouen. The quote above suggests that Mydorge was still working on applications of the sine law of refraction to lens theory ("we engage ourselves [for] placing the theory into practice"). I would therefore place the the *terminus ante quem* at the beginning of 1628.

Another interesting letter in the correspondence of Mersenne concerns a letter by Cornier to Mersenne 18 Aug [1625] writing from Rouen (translation by Shea):

---

[27] Mydorge (1631: 157–8).

[28] Schuster (2013: 186).

[29] CM (I, 404).

[30] Schuster (2013, 188).

[31] Mydorge (1631).

> I do not know of any other means of making parabolic mirrors beyond those that you know of, especially since you have the paper of Mr. Mydorge who knows all that can be known on the matter. I can only tell you that Mr. Le Vasseur says that he has found an absolutely certain way *by the sinuses*. But I cannot say more since I do not yet know how he goes about it.

So, while the method 'by the sinuses' does not refer to the sine law but to a construction method for parabola, we now know that in the summer of 1626, the rumor circulated in Paris that the engineer and cartographer Guillaume Levasseur from Dieppe, but living at Rouen, had found a way to reliably construct parabolic mirrors "only using the sines". The report by Robert Cornier, also a resident of Rouen, may have set in motion a series of events which we conjecture to have ensued as follows. We must assume the rumor reached Mydorge, who devised twenty different geometrical ways to trace the parabola for his upcoming book, and hence also Descartes. The latter decides to investigate the role of the sines in image location in mirrors and lenses and resorts to his "only master in optics", Kepler whose *Paralipomena* he studied at the beginning of 1616. Intrigued by Kepler's working hypothesis on refraction he further explores the ratio which remains invariant for the three angles of incidence (which we will discuss in the next section). In other words, he just 'reads' the sine law from a published diagram in Kepler's work. By the time that Cornier writes to Mersenne that Descartes probably will never find the laws of refraction (16 March 1626), he most likely already knew the sine law.

Descartes can hardly believe that he derived the sine law by purely geometrical means and looks for a confirmation in three ways: (1) by the measurement of the angles of refraction from air to water, (2) by refraction in glass prisms, and (3) through the construction of 'exact' lenses. As for the first issue, it is remarkable that at that time Mersenne repeatedly asks Cornier for observational data on the refraction of light in water (March 1626). In his answer of 22 March 1626, Cornier sends Mersenne a table with his data. In the 1636 edition of the *Harmonie Universelle*, Mersenne publishes the table by Cornier, his own table of measurements with three kinds of water, and a third table by "one of the most excellent minds of the century", obviously referring to Descartes.[32] For the confirmation of the sine law in a prism, Descartes seeks the assistance of Mydorge. For the confirmation of the sine law in the construction of convex elliptic and hyperbolic glasses, he relies on Ferrier and Mydorge.

Now, completely convinced by the validity of the sine law, Descartes communicates the law to his closest friends, Mersenne and Isaac Beeckman in 1628. Beeckman reports on the sine law in his *Journaal* on 8 Oct 1628 with a figure resembling that by Kepler.[33] Mersenne publishes the sine law with a very similar figure in his *Harmonie Universelle*. It then took Descartes several more years to develop a dynamic physical model to fit the sine law of refraction into his corpuscular theory of light.

---

[32] Mersenne (1636, prop. 29: 65–6).

[33] De Waard, (1939, III: 97).

So, instead of considering the anaclastic as a path to discovery, in our view the solution to this problem was an application of the sine law, confirming its validity. Descartes went as far in this confirmation process as designing a machine for grinding hyperbolic lenses. The text by Mydorge copied by Mersenne and the reference quoted above suggests that Mydorge played an important role in solving the anaclastic problem.

I will now move to Kepler's failed attempt to formulate a quantitative law for refraction as his starting hypotheses and methodology was the basis for Descartes' discovery. Our claim simply is that Descartes succeeded to see what Kepler missed out.

7.5 Kepler Looking for Invariant Proportions in Geometrical Diagrams

We now come to the geometrical diagram which is crucial to our story. Kepler's interest in optics was aroused by the solar eclipse of 10 July 1600. He studied the works of Alhazen, Witelo and Pecham and conducted many experiments. His *Ad Vitellionem paralipomena*, published in 1604, was a *status questiones* of all current knowledge on human vision and optics correcting many of the anomalies and mistakes he had found with the Perspectivists. He spent considerable efforts in finding a quantitative law for refraction because of its importance in providing an exact measure of atmospheric refraction. Chapter 4, §2 is dedicated to the "Refutation of various authors' various ways of measuring refraction" where he introduced a "procedure for measuring, to combine both the density of the medium and the angle of incidence." The diagram in fourth chapter of the *Paralipomena* related to this procedure is shown in Fig. 7.8. In another procedure for studying the measure of refraction, presented in the same chapter, Kepler relied on the analogy between reflection in spherical mirrors and refraction. This idea was probably borrowed from Della Porta who, in Bk. 2 of *De Refractione*, presupposed and demonstrated geometrically the reciprocal relation between reflection and refraction.[34]

The drawing is both a representation of empirical data (exact and co-exact attributes) as well as an ingenious analysis of refraction. Line *CB* represents the surface of a refracting medium (water). The three lines originating in point *A* are incidence rays. The lines originating in *R* represent the measured angles of refraction corresponding with the angles of the lines originating from *A*. The fact that the refraction lines all meet in point *R* is very unusual. A representation like this is the main step in discovering the sine law. Line *ED* is the bottom of the vessel filled with water. Line *KL* is a construction based on the crossing point of the perpendicular lines *HN*, *IM* and *DL* with the refracted rays. The perpendicular was also called *cathetus* and was used by the Perspectivists for locating the reflected or refracted image. This

[34] This was pointed out to me by Yaakov Zik.

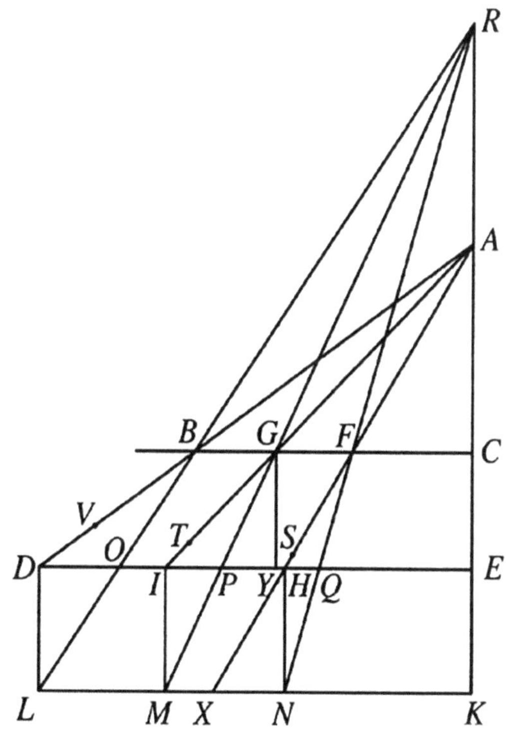

Fig. 7.8 Kepler's analysis of refraction in the *Paralipomena*, 1604, p. 65

clever representation allows to reason about refraction using geometrical knowledge about rectangular triangles. Given this drawing, Kepler now formulates a central hypothesis:

> Since density is obviously a cause of refraction, and refraction itself appears to be a kind of compression of light (i.e., towards the perpendicular), it comes to mind to ask whether the ratio of the media in the case of densities is the same as the ratio of the bottom of the spaces that light has entered into and strikes, first in an empty vessel, and then one filled with water.[35]

He then investigates several ratios between line segments in the drawing, which I have summarized in the Table 7.1. His formulation only refers to H01, the ratio between *EQ* and *EH*, but then extends the hypothesis to other ratios (numbered H02–H11). His belief is that one of these ratios corresponds to the "ratio of the media in the case of the densities", which we would now call the refraction index. After some pages of discussion he dismissed all eleven hypotheses as "refuted by experience, not to mention other procedures of enquiry".

[35] Kepler 1604, Ch. 4, § 2, Frish, 1859 Bd II, p. 182: "Cum ergo densitas plane sit in causa refractionum, et refractio ipsa compressio quaedam videtur lucis, utpote ad perpendicularem, subiit animum inquirere, an quae proportio mediorum causa densitatis eadem sit proportio fundi spatiorum, quae lux primum in vacuum vas, dein aqua superfusa, introgressa feriat." Translation by Donahue 2000, p. 102.

Table 7.1 Eleven hypotheses investigated by Kepler in his *Paralipomena*

H01	EQ/EH
H02	FQ/FH
H03	EQ^2/EH^2
H04	a.EQ/b.EH
H05	FHEC/FQEC
H06	EQ^3/EH^3
H07	IY/IP
H08	GC/IE
H09	CE/CK
H10	FH/FX
H11	CK/FX

We previously implemented Kepler's general hypothesis in a computational model.[36] In fact, the hypothesis he uses is more general than in the quotation above, because more ratios are investigated than those on the bottom of the vessel. A fair reformulation of Kepler's central hypothesis would be the following:

> The ratio of the optical densities of two media is proportional to some ratio of two line segments in the geometrical representation of a light ray traversing the two media, the first line segment related to the angle of incidence, the second to the angle of refraction.

Having implemented this central hypothesis in our computational model, it generated all of Kepler's hypotheses H01–H11, as well as many more. In fact, too many to be able to inspect manually. We notice from the table that, with exception of H08 and H11, all ratios of the two line segments have one point in common. If we add the restriction in the general hypothesis, that the line segments used in the ratios should start in the same point, the model generates only nine (linear) instances. These ratios cover all of Kepler's except the special case H07.

The next step is to either prove or disprove the generated hypotheses. Kepler's approach is to eliminate hypotheses by deduction or observation. Most hypotheses can be refuted deductively, by simple geometrical reasoning. For example, H11 states that the ratio *CK*/*FX* remains the same for varying angles of incidence. This is evidently not the case. As the angle of incidence increases, the length of *FX* will increase while *CK* remains the same. Therefore the ratio *CK*/*FX* decreases and hypothesis H11 is refuted. The tragedy of this case study is that Kepler succeeded in formulating both a suitable representation of the problem and the correct hypothesis that some geometrical proportion corresponds to the refraction index. He failed

[36] Heeffer (2003).

in identifying the invariant ratio. Both *FR*/*FA* and *FN*/*FH* correspond to the ratio of optical densities of the two media. These ratios can be proved to be invariant by geometrical reasoning. We can use line segment *FC* to establish a relation between the angles of incidence and refraction as it is the same side of the right-angled triangles *FRC* and *FAC*:

$$FA.\sin(FAC) = FR.\sin(FRC)$$
$$GA.\sin(GAC) = GR.\sin(GRC)$$
$$\frac{\sin(FAC)}{\sin(FRC)} = \frac{FR}{FA} = \frac{FN}{FH}$$

This ratio is a co-exact property of the geometrical diagram and remains invariant with respect to the angles of incidence. So is the ratio of the sines of angles *GAC* and *GRC* or *GR*/*GA* also equal to *FR*/*FA*. This unfortunate oversight was Kepler's failure in discovering the sine law.

7.6 The Balance Analogy

In Beeckman's *Journaal* we find a curious analogy when he reports about the sine law[37]:

> One of the angles of refraction being known, [Descartes] deduces the rest of them according to their sines: for as *ab* is to *hg*, thus *cd* is to *if*. For he assumes there is water below *st*, and the rays to be *aeg*, *cef*, and *they are seen to equilibrate themselves as the equal arms of some scale*, to the ends of which weights are hung, and the one in the water is lighter and raises the arms.

Malet writes that "it is difficult indeed to make physical sense of Beeckman's words and to reconcile them with sine law".[38] However, Schuster provides a comprehensive reconstruction of what Descartes had in mind, an exact model for the dynamics of refraction: "The conservation of the parallel component of the force of the light is modeled by the condition of equilibrium, which requires the equality of statical moments about the fulcrum".[39] It is also possible that the analogy was used in a more general way, employing the material model of a balance where the law of the lever keeps the product of the weights and arm lengths in balance. Equally, the products of the sines with the line segments, as in *FA*. sin (*FAC*) = *FR*. sin (*FRC*) remain invariant. Malet points out that Hérigone also adopts the balance analogy in

---

[37] De Waard (1939, III: 97, translation by Malet (1990), emphasis mine. Also compare with Fig. 7.7.

[38] Malet (1990).

[39] Schuster (2013: 200–203).

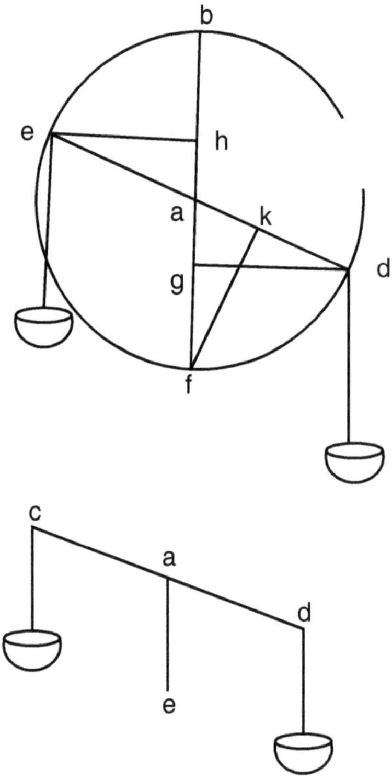

Fig. 7.9 Kepler's use of the balance analogy in the *Paralipomena*

his treatment of the sine law. However, Schuster, Malet and others seem to have overlooked an interesting passage in Kepler's *Paralipomena*[40] (Fig. 7.9):

> Even up to the present day it has not yet been adequately explained why here too there should occur a refraction, or rather (since this word is more appropriate to physical motion), a deflection, from the direct line of motion towards the line that is perpendicular to the surface (..) and because this whole matter depends upon the principle [ratio] of the balance, it should be derived from its source.[41]

Now, Kepler is using the analogy here in a more limited sense, explaining why light is refracted towards the perpendicular when moving to a denser medium. However, it is not difficult to see where Descartes found his inspiration for applying the balance analogy to the invariance of geometrical ratios. And we see in this another strong indication that Descartes discovered the sine law while studying the *Paralipomena*.

There is yet another interesting aspect about Kepler's figure: the outer points of the balance are encircled. None of the figures in Kepler's chapter 4 on refraction show a circle. Neither did Della Porta before him use circles for the analysis of refraction on a plane surface. But circles feature prominently in Descartes treatment

[40] Kepler (1604, prop. 20: 17–21).

[41] Kepler (1604: 18), translation by Donahue (2000).

of refraction. If you consider the circle as a unit circle with radius one, it becomes possible to *see the sines* in the figure. The sine of the opposite angle of a rectangular triangle is the ratio of the opposite line segment and the hypotenuse. However, in a unit circle the hypotenuse is 1 and thus the sine equals the opposite line segment. If we go back to Kepler's figure shown in Fig. 7.8 and one imagines unit circles with radius *AF* and *RF you see* that the sines of angles *FAC* and *FRC* are equal and are shown by the line segment *FC*.

If we now look back to what Mydorge did in the first proposition of his small treatise on refraction copied by Mersenne (see Fig. 7.5), we here discern the two semi-circles with different radius to be imagined in Kepler's diagram. Mydorge considers the angle with the cathetus, thus *CEF*, as the angle of incidence and the angle *DEG* as the angle on refraction, just as Kepler did. What Mydorge is presenting here is a method to construct the refracted ray for any angle of incidence given one incidence (*FE*) and one refracted ray (*EG*). In the terminology we have introduced earlier, his method depends on the invariance of co-exact attributes, while we mentally change the angle of incidence. Schuster insists on a cosecant version preceding the sine law formulation, but there is no reason to assume cosecants in propositions I and II, neither are cosecants mentioned by Mydorge. Instead, it can clearly be seen that the construction of Fig. 7.5 depends on the sine version. Mydorge concludes the first proposition with "leurs angles d'inclinations et de refraction doivent estre comparez ensemble suivant la raison de la droite *GE* à la droite *EF*". In a footnote, De Waard rightly notes that Mydorge here means the sines of the angles. Schuster however understands the conclusion of Proposition I as follows[42]:

> Mydorge observes that the law is given here as a law of cosecants. That is taking the first ray $\dfrac{\csc(i)}{\csc(r)} = \dfrac{R_1 / OF}{R_2 / OI}$ [with R_1 being *EF*, R_2 is *EG* and *O* being on the cathetus] since OF=OI, the cosecants are as the radius of upper semi-circle to the radius of lower semi-circle.

Consequently, he considers the unequal radius method as a cosecant version of the law of refraction, on which his further analysis is based. However, Mydorge does not use the term 'cosecant' at all and one can therefore equally state:

$$\frac{\sin(i)}{\sin(r)} = \frac{OF / R_1}{OI / R_2}$$

and interpret proposition I as the sine version, completely in accordance with the text, as endorsed by De Waard. Now, it becomes easy to see that the equal ratios *FR*:*FA*=*FN*:*FH* in Kepler's diagram indeed correspond to the ratios *EF*:*EG*=*EH*:*EN* in Mydorge's diagram, being the ratios of the sines of the angles of incidence and refraction. That is why we claim that Descartes read the sine law from Kepler's figure. In Proposition II, Mydorge applies the unequal radius method to the location

[42] Schuster (2013: 186).

of the refracted ray in a refractometer (as in the second diagram of Fig. 7.7). Schuster's heading discussing proposition II in Appendix I reads "Cosecants Not Sines". However, since the radius form of the law *is* the sine form, there is again no reason to consider proposition II a cosecant version. In a third and fourth proposition, Mydorge applies the sine law for refraction on hyperbolic and elliptic glass surfaces where he refers to sines explicitly. These are all confirmations of the sine law and the applications which Mydorge proudly claims to be his own inventions.[43]

7.7 Della Porta's Treatment of Refraction

The sine law of refraction was independently discovered by Thomas Harriot, Willebrord Snell and Descartes within a period of less than 30 years. The empirical data listing the angles of refraction for different angles of incidence for transition of air to glass, air to water and water to glass date back to Ptolemy's *Optics*. So, why then after so many centuries suddenly several natural philosophers came to derive the same law? Obviously, there was a body of shared knowledge present at the end of the sixteenth century, which made it possible to arrive at the sine law by geometrical reasoning. This understanding can best be summarized by a paraphrase of Kepler's general hypothesis: "the ratio of the optical densities of two media is proportional to some ratio of two line segments in the geometrical representation of a light ray traversing the two media". Kepler explored his hypothesis by means of geometrical diagram showing three lines with different angles of incidence and the corresponding co-exact lines of refraction. However, Kepler was not the first to do so. A very similar diagram appears in Della Porta's *De refractione optices*, (see Fig. 7.10).[44]

The context in which the diagram is used is different from Kepler's. Della Porta uses the diagram in a proposition which argues that things will appear higher in water the more distant they are from the eye. He shows three angles of incidence *AH*, *AR* and *AS* and the corresponding angles of refraction *OC*, *PD* and *QE*. In his explanation he also refers to point *G* which is not shown in the figure but is understood to be the refraction point where line *OC* crosses the surface *FK*. The correspondence with Kepler's figure is striking. All three lines *AH*, *AR* and *AS* start from a single point on the normal. The only difference with Kepler's diagram is that the crossing of the refracting lines with the normal are not shown. Kepler refers several

[43] The manuscript BNF NAF 5176, from which De Waard fabricated the "letter" from Mydorge to Mersenne contains at least two additional propositions which should have been included in his transcription. An interesting proposition (unnumbered, f. 28r) also concerns an application of the sine law to lens theory, in particular the focusing of parallel rays into a single point by use of a plano-convex (be it spherical or hyperbolic) lens. Descartes uses such a lens in his seventh discourse (AT, VI, 161).

[44] Della Porta (1593, Bk. 1, prop. 8).

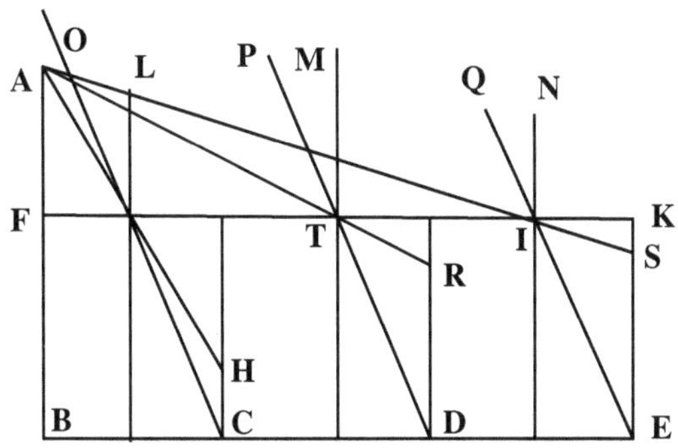

Fig. 7.10 Della Porta's figure the *De refraction optices*

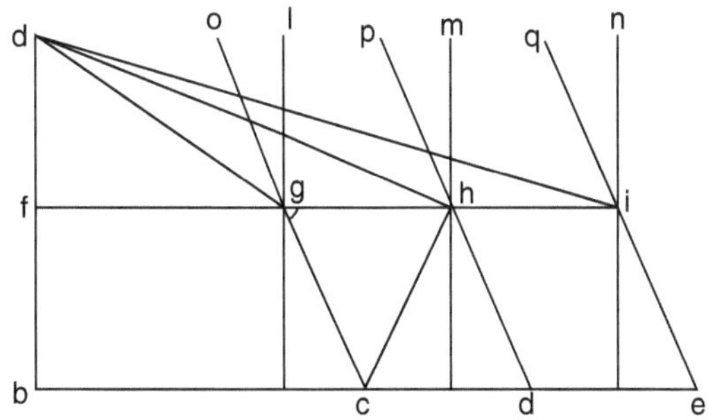

Fig. 7.11 Witelo's figure from the Basel edition, bk. 10, prop. 14, p. 416

times to Della Porta in his *Paralipomena* but they can all be determined to refer to the *Magia naturalis*. In fact, Kepler complains that "one Baptista Porta professed to have given an account in optics which I have hitherto sought in vain from booksellers", which is probably referring to *De refractione*. Some scholars have doubted his honesty in this respect, but the similarity of the diagrams can be explained by a diagram from Witelo which Kepler certainly knew (see Fig. 7.11).

Witelo's figure concerns the same proposition as Della Porta and uses the same numbering, showing point *G* but using *H* instead of *T*. But Della Porta criticizes Witelo for reaching the wrong conclusion that "sed angulus ogl & phm sunt aequales" which would imply that the angles of refraction are all the same while the angles of incidence differ. For the modern reader it is difficult to understand how Witelo could be so wrong. At the time of Kepler, the most important source for refraction tables was Witelo. The angles shown in the diagram are not consistent with the empirical data collected from his own experiments. This leads us to an important conclusion: Witelo did not respect the co-exact properties of optical diagrams. In arguing that the ratios *C*:*D*:*E* should be different from *G*:*T*:*I* Della Porta did account for the co-exact properties of the angles of refraction. Also Maurolyco in his *Photismi de lumine et umbra ad Diaphana* (published posthumously only in 1611) did not account for co-exact properties, claiming that there is a constant proportion of 8:3 between the angles of incidence and refraction. So, with Della Porta's

De refractione an important methodological step was taken: when reasoning with geometrical diagrams in optics, one has to account for the co-exact properties of the refracting lines. Kepler took the principle one step further, looking for invariances in these co-exact properties.

7.8 Epilogue: The Not-So-Exact Properties of Geometrical Diagrams

We have argued throughout this paper that one can arrive and that some did arrive at the sine law of refraction by geometrical reasoning when respecting the co-exact properties determined by the laws of dioptrics. If it is known of a geometrical diagram that a light ray enters the surface of water from air, the angle of refraction is co-exact in respect to the angle of incidence, determined by refractive index of water. These data are known from the refraction tables from Ptolemy and Witelo. In a more abstract setting, when the media are not known, the angles of refraction are still co-exact with respect to any given angle of incidence and corresponding angle of refraction. Mydorge's construction of Fig. 7.5 depends on these co-exact properties. However, if one inspects carefully the diagrams in the books of Della Porta, Kepler and Descartes, they appear not to follow these principles. Figure 7.10 is taken directly from the book published by Della Porta in Naples in 1593. Even while Della Porta criticizes Witelo for not respecting the correspondence between the angles of incidence and angles of refraction, he reproduces the defected figure, showing the three refracted rays with the same angle. Figure 7.12 shows the actual diagram from the Frankfurt edition of 1604.

If we measure the angles of incidence we find 28°, 40° and 50° respectively. Most likely, the intention was to show the entries in Witelo's refraction table for air to water, with incidence angles 30°, 40° and 50° which would correspond with the angles of refraction of 22.5°, 29° and 35°. However, we measure 10°, 22° and 29°, implying a refractive index of 1.76, 1.72 and 1.60 respectively. How should this be interpreted?

Since Della Porta was aware that the angles of refraction in Witelo's diagram do not match the angles of incidence, showing the wrong angles here had an argumentative function, but he did not take the effort to show a corrected figure. In Kepler's case we must assume that, as an able draftsmen proud of his skills in geometrical diagrams, he had the angles right in his manuscript notes. The geometrical drawings in Kepler's edition are woodcuts crafted by the printer or an engraver who is unlikely to have understood the restrictions posed by the laws of dioptrics.

Similar inaccuracies are to be found in Descartes's discourse on optics (see Fig. 7.13). Descartes was much less of a draftsmen than "his master in optics", Kepler. Descartes intended to show that the distance between *HB* and *FE* is twice that of *AH*, but the printer "misunderstood" the instructions. In contrast with talented draftsmen like Mydorge, Harriot and Newton, Descartes was very poor in drawing and was well aware of it. While getting help from Mersenne and Van

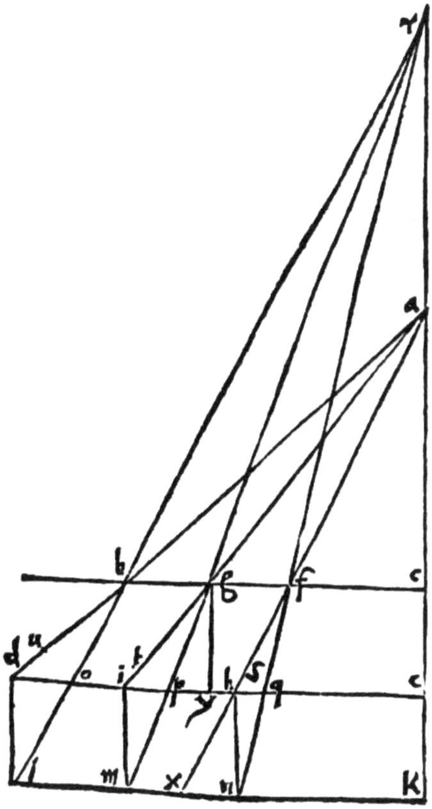

Fig. 7.12 Kepler's figure as it appears in the Frankfurt edition of 1604

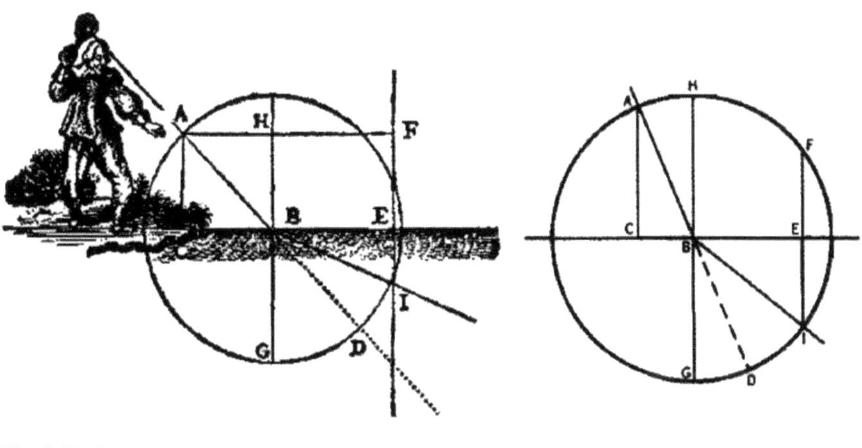

Fig. 7.13 Descartes' diagram of refraction as it appeared in his *Discourse* (*left*) and how it should appear with the correct proportions (*right*)

Schooten, the final woodcuts in the *Discourse* contained several errors. Hobbes took some pleasure in pointing out to Mersenne the misrepresentation of ratios in Fig. 17 (here shown as Fig. 7.13) of the *Optics*.

7.9 Conclusion

Starting from a short survey on reasoning with geometrical diagrams from philosophy of mathematical practice, we have identified some aspects crucial for our discussion of some optical diagrams: (1) the interpretation of geometrical diagrams is not limited to mere visual aspects but also implies kinetic mental operations such as change of angles, and (2) proofs often depend on the invariance co-exact attributes, which are those conditions that remain unaffected by some range of continuous variation within a specified diagram. In dioptrics, we find additional constraints that lead to invariances of co-exact attributes. Given the media separating a refractive surface, the angles of refraction are co-exact in respect to the angles of incidence or the angles of refraction are co-exact with respect to any given angle of incidence with a corresponding angle of refraction. We have demonstrated that by respecting the co-exact properties of optical diagrams the determination of an invariance can lead to the discovery of the sine law. Kepler's general hypothesis depended on the invariance of co-exact properties and while he failed to identify the correct ratio it can lead to the formulation of the sine law. We also showed that Mydorge's construction method also depends on the invariance of co-exact properties and we proposed the theory that Descartes discovered the sine law following Kepler's method. Although, diagrams simultaneously depicting different angles of incidence and refraction go back to Witelo, it was Della Porta who first insisted on respecting the co-exact properties.

In our argumentation, we have used modern concepts describing reasoning with geometrical diagrams, but these reflect specific procedures on lettered diagrams shared and understood in the Euclidean and Perspectivist traditions as shown in recent scholarly work.[45] One could object that we have lifted one specific diagram out of its specific textual context, but we believe this was done by participants in the historical process as well. Diagrams were taken out of a text and used as instruments almost in the same way as diopters and astrolabes. Even diagrams of the very instrument used to measure refraction became itself instruments for reasoning about refraction.

Acknowledgments This research paper resulted from the research project 3G002713 funded by the Flemish Research Foundation (FWO Vlaanderen), titled *A Study on Material Models from Engineering and Technology used in Discovery, Explanation and Negotiation in Early-Modern Philosophical Debates*. The paper benefitted from critical comments and suggestions by Maarten

[45] Netz (1999) and Manders (2008).

Van Dyck, Boris Demarest, Arianna Borrelli, A. Mark Smith, Yaakov Zik, John Schuster and proofreading by Saskia Willaert.

References

Ariew, R. (tr.) 2000. René Descartes. *Philosophical essays and correspondence*, Indianapolis: Hacket Publishing.

Buchdahl, G. 1969. *Metaphysics and the Philosophy of Science*. Cambridge MA: MIT Press.

Della Porta, G. 1593. *De refractione optices parte: libri novem*. Neapoli: Ex officina Horatii Salviani, apud Jo. Jacobum Carlinum, & Antonium Pacem.

Descartes, R. 1897-1913. *Oeuvres De Descartes*, 12 vols., eds. by Charles Adam and Paul Tannery, Paris: Librairie Philosophique J. Vrin, 1983 [AT].

Donahue, W. 2000. *Kepler's Optics*. Santa Fe: Green Lion Press.

Frish, C. 1858. *Joannis Kepleri Astronomi opera omnia*. Francofurti: A.M., Heyder & Zimmer.

Giardino, V. 2017. Diagrammatic reasoning in Mathematics. In *Springer Handbook of Model Based Science*, eds. Lorenzo Magnani & Tommaso Bertolotti, New York: Springer (forthcoming).

Heeffer, A. 2003. Kepler's near discovery of the sine law: A qualitative computational model", In *Computer Modeling of Scientific Reasoning*, eds. Claudio Delrieux & Javier Legris, Bahia Blanca, Argentinia: Universidad Nacional Del Sur: EDIUNS, 93–102.

———. 2006. The Logic of Disguise: Descartes' Discovery of the Sine Law. *Historia Scientiarum* 16(2): 144–165.

Hérigone, P. 1634. *Cursus mathematicus: nova, brevi, et clara methodo demonstratus ... Cours mathematique, demonstré d'une nouvelle, briefve, et claire methode*. Paris: Chez Henri Le Gras.

Kepler, J. 1604. *Ad Vitellionem paralipomena quibus Astronomiæ pars optica traditvr: potissimum de artificiosa observatione et æstimatione diametrorvm deliquiorum[que] solis et lunæ, cvm exemplis insignivm eclipsivm : Habes hoc libro, lector, inter alia multa noua, tractatum luculentum de modo visionis, & humorum oculi vsu, contra opticos & anatomicos*. Francofvrti: Apud Claudium Marnium & Hæredes Ioannis Aubrii.

Kirsh, D., and P. Maglio. 1994. On Distinguishing Epistemic from Pragmatic Action. *Cognitive Science* 18: 513–549.

Kramer, P. 1882. Descartes und das Brechungsgesetz des Lichtes. *Abhandlungen zur Geschichte der Mathematik* 4: 233–278.

Mach, E. 1926. *The principles of physical optics: an historical and philosophical treatment*. New York: Dover.

Malet, A. 1990. Gregorie, Descartes, Kepler, and the law of refraction. *Archives Internationales D'histoire Des Sciences*. 40: 278–304.

Manders, K. 2008. The Euclidean diagram. In *Philosophy of Mathematical Practice*, ed. Paolo Mancosu, 112–183. Oxford: Clarendon Press.

Mersenne, M. 1636. *Harmonie universelle contenant la théorie et la pratique de la musique: où il est traité de la nature des sons, & des mouuemens, des consonances, des dissonances, des genres, des modes, de la composition, de la voix, des chants, & de toutes sortes d'instrumens harmoniques*. A Paris: Chez Sebastien Cramoisy.

Milhaud, G. 1921. *Descartes savant*. Paris: Felix Alcan.

Mumma, J. 2010. Proofs, Pictures and Euclid. *Synthese* 175(2): 255–287.

Mydorge, C. 1630. *Examen du livre des Recreations mathematiques et de ses problemes en geometrie, mechanique, optique, & catoptrique. Où sont aussi discutées & restablies plusieurs experiences physiques y proposées. Par Claude Mydorge, escuyer, sieur de La Maillarde, conseiller du Roy, & tresorier general de France en Picardie.* A Paris: chez Rolet Boutonné, au Palais, en la premiere boutique de la gallerie des prisonniers.

Mydorge, C. 1631. *Clavdii Mydorgii patricii Parisini Prodromi catoptricorvm et dioptricorvm: sive, Conicorvm operis ad abdita radii reflexi et refracti mysteria praeuij & facem praeferentis: libri primvs et secvndvs* : D.A.L.G. Parisiis: Ex Typographia I. Dedin.

Netz, R. 1999. *The shaping of deduction in Greek mathematics: a study of cognitive history.* Cambridge: Cambridge University Press.

Sabra, A.I. 1967. *Theories of light from Descartes to Newton.* London: Oldbourne.

Schuster, J.A. 1978. *Descartes and the Scientific Revolution 1618–1634: An Interpretalion.* PhD thesis Princeton University. Ann Arbor. Mi: University Microfilms International.

———. 2013. *Descartes-Agonistes: Physico-Mathematics, Method and Corpuscular-Mechanism. Studies in History and Philosophy of Science.* New York: Springer.

Shea, W.R. 1991. *The magic of numbers and motion: the scientific career of René Descartes.* Canton, MA: Science History Publications, U.S.A.

Smith, A.M. 1996. *Ptolemy's theory of visual perception: an English translation of the Optics.* Philadelphia: American Philosophical Society.

Waard, C. De. 1939. *Journal tenu par Isaac Beeckman de 1604 à 1634.* La Haye: M. Nijhoff.

———, 1986. *Correspondance du P. Marin Mersenne, religieux minime.* Paris: Éditions du Centre national de la recherche scientifique, 16 vols [CM].

Chapter 8
Francesco Maurolico, Giambattista Della Porta and Their Theories on Refraction

Riccardo Bellé

8.1 Introduction

For a reassessment of Della Porta's optics a broad view is needed. For a better understanding of the similarities and the peculiarities of Della Porta's approach in comparison with other authors of the same period, we are going to deal with Maurolico as one of the best candidates.

Maurolico and Della Porta share some common features: both of them proposed a law of refraction, studied behaviour of light passing through glass spheres and discussed convex and concave lenses, trying to connect the known, or newly discovered, optical effects to the theory previously expounded. What makes the comparison even more interesting is that their approaches were very different and complementary to each other: Maurolico was a mathematician who devoted his energy to the restoration of ancient Greek geometry, while Della Porta was a very eclectic polymath interested in experimental and astounding phenomena, but he was not neglecting reconstruction and interpretation of ancient scientific works.

According to Rose (1975, p. 159), who wrote a whole chapter on *Maurolico and the Renaissance of Greek Mathematics*, Maurolico was "perhaps the greatest geometer of the sixteenth century". His technique consisted in correcting the works of ancient authors, but also of writing texts with new ideas and demonstrations, departing from contents of classical works. But, unlike Della Porta, whose name was known all over Europe with many editions of his *Magia naturalis*, Maurolico's most important works were printed "only after their novel ideas had been re-discovered independently and published by other mathematicians" (Rose 1975, p. 160).

R. Bellé (✉)
Liceo Scientifico "Enrico Fermi", via E. Fermi 2, 54100 Massa, Italy
e-mail: rickbel@tiscali.it

A. Borrelli et al. (eds.), *The Optics of Giambattista Della Porta (ca. 1535–1615): A Reassessment*, Archimedes 44, DOI 10.1007/978-3-319-50215-1_8

Della Porta was famous for his wide range of knowledge and the importance he gave to the observation and the description of natural phenomena. He always tried to find the unity in nature, as Clubb (1965, p. 55) wrote:

> Illustrating the Renaissance search for unity in nature, all his works take for granted a universal plan which expresses itself in analogies among all forms of existence: between men and animals in the *Physiognomonia* and *Chirofisionomia*, between animals and plants in *Phytognomonica*. But utility was always Della Porta's highest aim; every book of the *Magia* emphasises the practical application of the secrets of nature revealed therein.

Also in optics Della Porta based his most valuable results on analogy more than on a rigid and clear deductive structure. Nonetheless his experiments and descriptions must be taken seriously, especially in the case of *De refractione* which represents, in some passages, a progress in comparison to what is found (or not found) in other texts of the same period.

In Sect. 8.2 of this paper a concise description of the relation between mathematics and its use in discussing physical phenomena in the sixteenth century is found. Then in Sects. 8.3, 8.4, 8.5, and 8.6 follows an account of Maurolico's works in optics and a presentation of his most relevant results, with a brief discussion of the influence on subsequent studies (if any). Next in Sects. 8.7, 8.8, and 8.9 Della Porta is examined, following more or less the same path of exposition: works composed in the field, evolution of approach and discussion of results. For both authors, the focus is on refraction of light passing through glass spheres and lenses.

In final Sect. 8.10, I will try to make a comparison between the two approaches, showing differences and similarities, concluding with the influence and the relevance of both points of view for seventeenth century optics. I argue that the great amount of results obtained by seventeenth century scholars in this field derived from a fruitful melting of these two different ways to deal with optical problems:

1. a geometrical way, interested less in phenomena and experiments and more in demonstrations mathematically tenable;
2. an approach not rigorous, but as much useful, aimed to find explanations of phenomena and to present them with geometrical appearance, although not in a rigorous manner, giving a structure to a collection of observations and experiments which were not previously organised in systematic way.

8.2 Mathematization in the Sixteenth Century

The idea that some physical properties of bodies could be treated in the setting of the theory of proportion is a commonplace used by the majority of the comments to the Euclidean definition of ratio. In the sixteenth century, however, there are the first significant attempts for translating in a mathematical language a wide range of physical phenomena: the theory of motion by Galilei, the static *momentum* and the law of refraction by Maurolico, the model of the specific gravity in Ghetaldi and Galilei.[1]

This is the context in which to situate also the optical studies of the sixteenth century. In those years the first attempts to lay the foundation of mixed sciences were found, through rigorous mathematical demonstrations, combined with observations and experiments. This approach provided sometimes an exact, although qualitative, description of natural phenomena together with an account of how to reproduce expected effects. In other cases, the observed phenomena are justified (or explained) by logical deductions from a set of suppositions, in a more or less clear deductive "Euclidean" structure. Maurolico and Della Porta stand on opposite sides: for Maurolico the focus is on mathematics, but he will not neglect experiments and simple descriptions of phenomena. For Della Porta the contrary is true: experiment and its description is at the center, but in *De refractione* he tried, not always successfully, to give a geometrical structure to his reasoning.

8.2.1 Physical Properties and Proportion Theory

A quantitative view of natural phenomena is achieved through a description of the sensible world using mathematical language. The main tool at disposal in the sixteenth century is the Euclidean theory of proportion. However, Della Porta's *De refractione* demonstrates that another approach is possible. Della Porta, in fact, suggested an investigation of nature with different research tools.

Every attempt to elaborate a geometric model of physical quantities must employ on the one hand the rigid language of the proportion theory and face on the other the problem to explain new concepts with old tools. But this was not the only problem. Even in the the sixteenth century, the proportion theory was not fully and correctly understood by every scholar. In fact, Campanus' translation, one of the most used and known version of Euclid's *Elements* and the first to be printed, presented the definition of proportionality of four quantities as the proportionality of their equimultiples, which is an absurd.[2] Campanus' translation (quantities are proportional if

[1] One of the most influential commentaries on Euclid of the period was (Clavius 1574), but already Campanus maintained that proportion is found not only in quantities (geometrical or numerical), but also in weights, forces and sounds: "Non enim solum in quantitatibus reperitur proportio, sed in ponderibus, potentiis et sonis" (Campanus 1546, p. 103). Quite representative of this spirit was the work by Cardanus *Opus novum de proportionibus numerorum, motuum, ponderum, sonorum aliarumque rerum mensurandarum* (1570). On the model of specific gravity see (Napolitani 1988). For Galilei, whose model is found in the short text *De motu equabili* (1609–1612), see (Giusti 1986) and (Giusti 1993). For Maurolico's statics see (Giusti 2001). Maurolico's refraction law will be treated in this contribution.

[2] "Quantitates quae dicuntur esse secundum proportionem unam, prima ad secundam et tertia ad quartam, sunt quarum primae et tertiae multiplices aequales, multiplicibus secundae et quartae aequalibus fuerint similes, vel additione, vel diminutione, vel aequalitate eodem ordine sumptae", def. 6, bk. V, (Campanus 1546, p. 105). "Magnitudes are said to be in proportion, the first to the second and the third to the fourth when the equimultiples of the first and the third are, according

and only if so are their multiples) gave to the reader a meaningless definition, essentially a circular reasoning. Nonetheless Maurolico succeeded in composing a completely satisfying demonstration, as the Sect. 8.5 will show. First of all, Maurolician optical studies are described.

8.3 Maurolico's Optical Studies: An Overview

Francesco Maurolico was born in Messina in 1494, in the middle of the renewed interest in the recovery and restoration of classical works, in particular, in the field of mathematics and science. This process was not straightforward for at least two reasons: first, the texts are often transmitted with lacunae or errors, collected during the centuries; second, it was necessary to reconstruct the scientific background in which the works were written, in order to understand theories and problems handed down. When sixteenth century scholars found unclear passages in the texts, they felt the need to integrate and explain them.

As a result of this process, the efforts to restore the classical Greek geometry laid the basis for a new science. In doing this, two roads were at their disposal: the first one used philology, with special attention to be faithful to the original work; the other, instead, made use of mathematical reasoning to guess the lost passages or even to reconstruct entire texts only on the basis of their theoretical content. Clearly these two aspects were not completely distinct: after all they share the same goals and complement each other. Maurolico chose the mathematical way (Rose 1975, p. 166):

> Maurolico is concerned purely with the mathematical aspects of the texts; ...In principle there was nothing wrong with correcting classical treatises, as long as it be done by an expert mathematician. Nonetheless, Maurolico was certainly the most radical of Renaissance mathematicians in this respect.

Optics was one of Maurolico's interests from the very beginning of his mathematical career: *Photismi* and *Diaphana*, dated respectively 1521 and 1523, are among his earliest scientific works. His interest in optics is reflected by the preface to *Grammaticorum rudimentorum libelli sex* (1528, ff 7r-v), where Maurolico outlines his scientific program[3]:

---

the to decrease or the increase or the equality, similar to the equimultiples of the second and the fourth taken in the corresponding order". Clavius wrote that Campanus' interpretation is absurd because it is simply a *petitio principii*: "si ita intelligatur definitio, Euclidem idem per idem definire" (Clavius 1574, p. 156).

[3]"Nam universa *Elementorum* Euclidis volumina et eiusdem *Optica* et *Catoptrica* ...nullo praeeunte praeceptore, per memetipsum intellexi. Alii due [*scil.* libelli adiicientur] *de Photismis* unus, alter *de Diaphanis*. In illo, praeter caetera, patescet cur solaris radius per qualecumque foramen transmissus in circularem redigatur formam; in hoc ratio rotunditatis et colorum iridis aperietur. Quorum utrunque fuit Ioanni vulgatae *Perspectivae* authori incognitum".

> Without teacher, by myself, I studied all the books of Euclid's *Elements* and his *Optics* and *Catoptrics* …and [I wrote] two other books, *Photismi* and *Diaphana*. In the former, it is explained why solar rays, passing through a hole of whatsoever shape, obtain a circular shape; in the latter, I presented the explanation of the round shape and the colors of the rainbow. Both of these topics were unknown to John Pecham, author of *Perspectiva communis*.

Later in his career he will modify and expand his optical studies and increase his optical sources. Over time, he produced an increasing number of original works. In the years 1553–1554, for example, Maurolico put his hand to the text of *Diaphana*, revising it extensively and adding a whole new section devoted to the physiology of vision. Finally, in 1568, *Problemata ad perspectivam et iridem pertinentia* appeared as an appendix to *Aristotelis problemata mechanica*.[4]

Unfortunately, of the wide array of known works on optics by indirect sources, only four remain at present. They were published all together in Naples in 1611: *Photismi*, *Diaphana*, *Problemata ad perspectivam et iridem pertinentia*, and a very short text, a sort of appendix to *Photismi*: *De erroribus speculorum*. Maurolico's interest in optical sciences is quite extensive and covers nearly his entire career, from 1522 to 1568, a few years before his death in 1575.[5]

Maurolico's optical works raised also an interest in the twentieth century with an English translation (Crew 1940) and an Italian partial translation (Ronchi 1968).[6]

A critical edition of Maurolico's optical works is forthcoming, edited by Riccardo Bellé and Ken'ichi Takahashi, as volume 10 of the *Edizione Nazionale dell'Opera matematica di Francesco Maurolico*, promoted by the Italian Ministry of Culture.[7]

---

[4] For reference and more information on the complete Maurolician *Corpus* in optics see (Bellé 2006a).

[5] Maurolico's works on optics are extant in four manuscripts and two printed editions. Only the list of manuscripts and printed editions is give here, see (Bellé 2006a) for detailed information. Paris, Bibliothèque Nationale, *Lat.* 7249, is an autograph manuscript containing only *Diaphana*; Lucca, Biblioteca Statale, ms. 2080, is a copy of the seventeenth century containing all the works except *Problemata*. *Problemata* are found in *Fondo Curia* 2052, Archivio della Pontificia Università Gregoriana, a manuscript written by C. Clavius. Finally, Hamburg Staats und Universitätsbibliothek, *Cod. Math.* 483 is a later copy which includes only the enunciation of theorems. The first printed edition is *Photismi de lumine et umbra, Diaphanorum partes, seu libri tres, Problemata ad perspectivam et iridem pertinentia*, Napoli, *Ex Typographia Tarquinij Longi*, 1611. Two years later there was a second edition, with some variations, in Lyon, with title *Theoremata de lumine et umbra* instead of *Photismi*.

[6] Ronchi translated only the first part (theorems I–III) and the last part (theorems XXX–XXXV) of *Photismi*, *De erroribus speculorum* and the third part of *Diaphana*. In the same book some passages from Della Porta's *Magia naturalis*, *De refractione* and *De telescopio* are translated.

[7] For an electronic version of our edition and further information, see www.maurolico.it.

8.4 Maurolico's *Diaphana*

Diaphana deals with refraction. I will focus on this work in comparison with *De refractione* in order to show similarities and differences in content and approach between Maurolico and Della Porta.

It is divided into three parts: "Pars prima" is devoted to vision caused by deviated rays, the law of refraction and the behaviour of light through transparent spheres; "Pars secunda" to the rainbow; "Pars tertia" to the structure of the eye and the mechanism of vision. This book had a first version, dated January 1523, and a subsequent revision thirty years later (1553–1554). In this second phase, it underwent a major overhaul regarding several theorems, mostly in the second half, and the addition of important topics. The entire third part, for example, dates from this period.

The text begins with a "definition", which merely states a well-known effect of refraction, taken from Euclid's *Catoptrics* as Maurolico himself explicitly states.[8]

After the definition, there are four suppositions and 24 theorems. The first group deals with vision by deviated rays (theorems I–IX). Maurolico shows, for example, that an object within a transparent, plane medium appears larger and closer to the surface (theorem I); he also studies refraction through a transparent sheet with flat and parallel surfaces (theorem V) and through prisms (theorems IV and VI); finally he proves that an object partially immersed in a transparent substance appears broken.

The following group consists of two theorems and several corollaries, and it is devoted to the law of refraction. The most important result is presented in theorem X: "Inclination angles are proportional to fraction angles".[9] This theorem is taken under scrutiny in the following section.

The last group, which includes theorems XII–XXIV, deals almost exclusively with the behaviour of light through transparent spheres. Maurolico studies the path of rays in various situations: for example, within a sphere, when the incoming rays are parallel (theorem XII) or the behaviour of the outgoing rays if they are parallel inside the sphere (theorem XVIII). These considerations lead him to speak of lenses: in theorem XXIII, for example, he shows that the image produced through a transparent sphere is reversed.[10] The first part of *Diaphana* concludes with the description of a burning lens.

The second part, as indicated by the title (*De Iride*), deals with the rainbow: its generation, the angle under which it is seen and the formation of colours. Secondary

---

[8] All the reference to *Diaphana* are given with paragraph number of our edition and with page number of (Maurolico 1611): "ut ait Euclides" (Diaph. § 1, p. 31).

[9] "Anguli inclinationum sunt fractionum angulis proportionales" (Diaph. § 33, p. 36). See note 14 for an explaination of the expression *fraction angle*.

[10] The result is incorrect and Maurolico's analysis is, in this case, unsatisfactory. Della Porta, on the contrary, will find correct results on this topic. See in particular Sect. 8.8.2.

rainbow is examined in a lengthy discussion, in particular concerning the angle under which the secondary rainbow is seen.[11]

The third part deals with five different but strictly related topics: anatomical structure of the eye; effects on rays of biconvex and biconcave lenses; mechanism of vision; causes of myopia and hyperopia and study of glasses for their correction[12]; discussion on the burning lens.

8.5 The Law of Refraction and Its Application

The Maurolician refraction law relies on an application of the Euclidean theory of proportion and, in particular, on the four suppositions that open the text[13]:

1. A ray incident perpendicularly upon a transparent surface proceeds in straight line; an oblique ray is bent towards the perpendicular.
2. Rays equally bent are equally refracted; more bent rays are more refracted.
3. If inclination angle is multiplied, the fraction angle is equally multiplied.[14]
4. The object is seen where the visual ray intersects the perpendicular to the plane surface of the transparent body.

In theorem X Maurolico tries to demonstrate that the inclination angles are proportional to the fraction angles. Here is how the demonstration proceeds.

In a plane *diaphanum* consider two oblique incident rays *CD* and *EF*. They are refracted towards the perpendicular lines *AK* and *BL* (to the point *G* and *H* respectively), according to the first supposition. We want to demonstrate that angle *DAK* is to angle *FBL* as angle *GAD* is to angle *HBF*.[15] Let rays *MAN* and *OBP* (refracted to point *Q* and *R* respectively) be such that *NAK* and *PBL* be whatever multiples of *DAK* and *FBL* respectively. Then, for third supposition, also *NAQ* and *PBR* will be multiples of *DAG* and *FBH* respectively. But, for the second supposition, if angle *NAK* is greater than angle *PBL*, also angle *NAQ* will be greater than angle *PBR*; and if lesser, lesser; and if equal, equal. So, for the definition of proportional magnitudes angle *DAK* is to angle *FBL* as *DAG* is to *FBH*, q. e. d. (Fig. 8.1)

---

[11] This part underwent many changes in the revision of the text dating from 1553–1554. For example, dealing with secondary rainbow, Maurolico changed the value of the angle under which it is seen: 55 degrees in the first version, 56 and one quarter degrees in the second one. For more information, see the Note to the Text of *Diaphana* in our edition www.maurolico.it/Maurolico/sezione.html?path=10.3.

[12] According to Lindberg (1984, p. 141): "Maurolico's most original contribution to visual theory".

[13] The translation is taken, with some variations, from (Crew 1940, pp. 49–50).

[14] As usual for the period, the *fraction* angle is the angle of deviation, that is the difference between inclination angle and what nowadays is called *refraction* angle.

[15] Angles *DAK* and *FBL* are equal to inclination angle, being vertical angles. Angles *GAD* and *HBF* are deviation angles, called fraction angles.

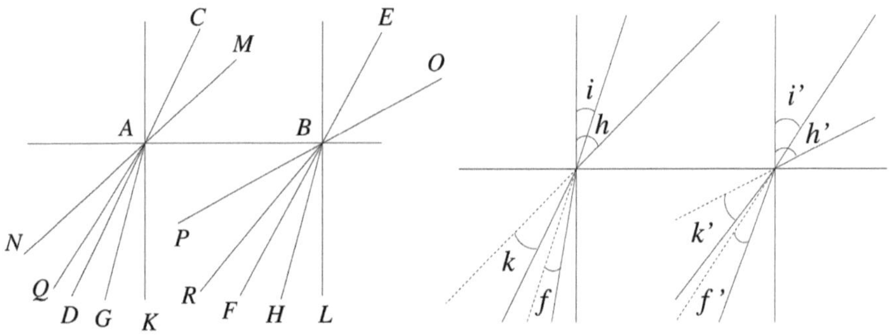

Fig. 8.1 Theorem X. Original and modern notation

Let's sketch Maurolico's reasoning in modern terms[16]:

1. He constructs the multiples h and h' of angles i ($h = ni$) and i' ($h' = mi'$), where n and m are natural numbers.
2. According to the third supposition, if the inclination angle is multiplied, the fraction angle is equally multiplied[17]; so angles k and k', fraction angles corresponding to the inclination angles h and h', are equally multiplied: $k = nf$ e $k' = mf'$.
3. The second supposition states: rays equally bent are equally refracted; more bent rays are more refracted.[18] So it implies these inequalities:

$$h > h' \Rightarrow k > k'$$
$$h = h' \Rightarrow k = k'$$
$$h < h' \Rightarrow k < k$$

4. Euclidean definition of proportionality (with point 3) gives $i : i' = f : f'$.

Maurolico understands very clearly the proportion model and how to choose the suppositions to demonstrate a law of proportionality. This is a quite natural choice in his period: the Euclidean theory of proportion was the only model at disposal of a mathematician. The case of refraction is quite simple, from this point of view. Maurolico had to deal only with magnitudes of the same kind: inclination and fraction angles.

But the question is how to build a geometrical model of physical properties of bodies. The Euclidean definition of ratio is (Heath 1956, p. 114): "Ratio is a sort of

[16] See Fig. 8.1, where f, k and f', k' are fraction angles corresponding to the inclination angles i,h and i', h'.

[17] "Multiplicato angulo inclinationis, angulum quoque fractionis aequaliter multiplicari" (Diaph. § 4, p. 31).

[18] "Radios aeque inclinatos aeque frangi, magis vero inclinatum, magis" (Diaph. § 3, p. 31).

relation between two quantities of the same kind in respect of size".[19] So, first of all, it is necessary to establish which properties of bodies can be considered "in respect of size". Maurolico gave a very clear answer to this problem in his *Prologi sive sermones quidam* (Maurolico 1968, p. 46), where he explains the model of mechanical momentum in statics[20]:

> Weight and lightness being respective qualities ...can be examined according to their quantity ...when a body is lighter than another then the second one is heavier than the first ... similarly the same voice is higher compared to a low voice, but lower compared to an high one. And as voices, so also the weight are compared according to proportion.

First of all it is necessary to establish which magnitudes can be considered from a quantitative point of view, the second step is then to build a demonstrative structure, borrowed from Euclidean theory, in order to find the relations between the different properties. In modern notation, Euclidean definition can be written:

$$A : B = C : D \Leftrightarrow \forall n, m \in N$$

$$nA > mB \Rightarrow nC > mD$$
$$na = mB \Rightarrow nC = mD$$
$$nA < mB \Rightarrow nC < mD$$

The definition itself inspires a relation based on comparison between magnitudes. In order to demonstrate that four magnitudes are proportional, it is necessary to make comparison between magnitudes and to build multiples of the magnitudes considered. So, the procedure we have seen "in action" in theorem X could be summarized in three steps:

1. choosing some physical-mathematical properties (quantitative) describing the phenomena[21];
2. developing the demonstrative structure of Euclidean theory of proportion, in order to obtain that the chosen physical properties are proportional each other;
3. applying the mathematical results obtained in step 2 to the physical world, for example to refraction between air and crystal, to get some numerical values.

As shown, Maurolico followed this path in building the demonstration of theorem X. Then he was immediately able to find the ratio between inclination angle and fraction angle (a corollary to theorem X). In fact, according to *alternando* rule,

[19] "Ratio est duarum magnitudinum eiusdem generis aliquatenus ad invicem quaedam habitudo" (Zamberti 1505, def. 3, bk. V).

[20] "Pondus quoque ac levitas cum respectivae qualitates sint ...ad quantitatem redigi possunt. Qualitas enim talis respectiva est: cum idem corpus alio respectu leve sit, alio autem grave ... Similiter vox eadem respectu vocis acutioris gravis est, respectu autem gravioris acuta. Sicut igitur voces inter se proportione, ita et pondera comparantur".

[21] In doing this, Maurolico clearly relied on what previous authors had done about the same subject.

$i{:}f{=}i'{:}f'$.[22] At this point Maurolico concluded that the angle of inclination has a fixed ratio with the fraction angle. Then, by an experiment with a crystal sphere, he determined this ratio equal to 2 + 2/3 (that is 8:3).[23] The fixed ratio refers to a given pair of media: in this case air and crystal. From Maurolico's law, it is possible to obtain a value of 1.6 for the "refractive index" of crystal.[24] Since refraction angle (r) is equal to the difference between incidence angle (i) and Maurolico's fraction angle (f) it is possible to write:

$$i:f=8:3 \qquad i:i-f=8:8-3 \qquad i:r=8:5 \qquad i:r=1.6$$

In Maurolico, a general analysis of different media is lacking. For example, Ptolemy and Witelo had different tables, showing different values.[25] Maurolico makes no mention of this question, and so his model, from the experimental point of view, is clearly defective. Maurolician law is nonetheless remarkable, considering the difficulties that Euclidean definition had posed at Maurolico's time and even thereafter. If one thinks of the circulation of Campanus' version of Euclid's *Elements*, which shows an incorrect and obscure translation, it is impossible not to appreciate Maurolico's mathematical exactness and clearness in applying Euclidean definition to construct the demonstration of theorem X of his *Diaphana*.

8.5.1 Back to Optics: Application of Theorem X

What is even more interesting is that Maurolico did not confine himself to finding the law of refraction but he used the newly established law to investigate the behaviour of light through glass spheres.[26]

Theorems XII–XXIV are devoted to this topic, covering a wide range of cases. My analysis will focus on theorem XII, with a brief mention to theorem XIX and its corollary, where a value for the focal point is found.

---

[22] In the proportion theory a *ratio* is a relation between two magnitudes of the same kind. In more complex models, like static momentum, specific gravity or theory of motion, which involve magnitudes of different kind, the *alternando* rule is not allowed.

[23] "Ergo et angulus inclinationis ad angulum suae fractionis semper unam servat rationem. Sicut experimento in crystallina sphaera probavimus. Estque dupla et 2/3 superpartiens" (Diaph. § 4, p. 31).

[24] I call here refractive index the fixed ratio between inclination and refraction angle, in analogy with the term used nowadays.

[25] In Ptolemy three tables are found: air to water, air to glass and water to glass, see (Smith 1996, pp. 233, 236 and 238). In Witelo also the reciprocal path is found, that is water to air, glass to air and glass to water. The values are erroneous, probably computed by a wrong application of reciprocal law, see (Grant 1974, p. 425).

[26] This application justifies, in part, the mentioned lack of interest for media other than glass.

Theorem XII deals with parallel rays incident upon a transparent sphere: the refracted rays, inside the sphere, will get closer each other, but with no intersection (Fig. 8.2).[27]

The demonstration is a *reductio ad absurdum*: if one ray were to intersect the other, then it would be impossible for the refraction law stated in theorem X to hold. For example, in the first examined case (Fig. 8.3), *EL* cannot reach point *B* (or intersect *AB*), because, if *EL* arrived at *B*, then the ratio $\sphericalangle FEN:\sphericalangle FEL$ would be equal to 2 (or even lesser than 2, when *EL* intersects *AB*). But this is impossible, being that $\sphericalangle FEN:\sphericalangle FEL=i:f$ and, according to the corollary of theorem X, the ratio between inclination angle and fraction angle equals 8:3 which is greater than 2. The other case, where neither of the two rays passes through the center of the sphere, is more complex but the same reductio holds (Fig. 8.4).[28]

In this theorem it is possible to see the law of refraction "being at work", used to deduce a theorem on natural phenomena; unfortunately the deduced consequence is not empirically correct. In nature, in fact, the rays intersect inside the sphere. I made

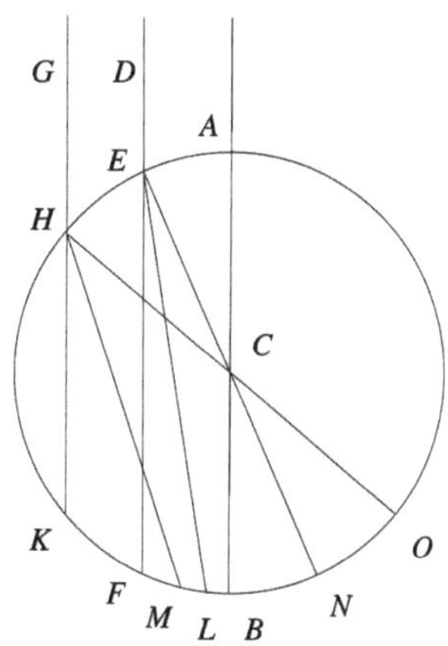

Fig. 8.2 Theorem XII The ray *AB* goes unrefracted. The ray *DE* is refracted towards *AB* (refracted ray *EL*), but doesn't intersect it. The ray *GH* is refracted towards *EL* (refracted ray *HM*) but without intersecting it

[27] "Given three parallel rays incident upon a transparent sphere, one passing through the center, the other two beyond the center; then the one passing through the center will maintain its direction in a straight line; the next ray will be refracted towards the first, but not sufficiently to intersect it; the remaining ray will also be refracted, but will never reach the other refracted ray" (Crew 1940, p. 58). "Transeuntibus per diaphanam sphaeram tribus parallelis radiis uno per centrum reliquis autem praeter centrum; qui per centrum in rectum transibit, at ei propior frangetur, ad ipsum tamen non perveniet. Reliquus quoque frangetur, sed alterum fractum nequaquam continget" (Diaph. § 41, p. 36).

[28] For a brief sketch of demonstration see Fig. 8.4. In the demonstration the arcs of the circumference and the corresponding angles are sometimes interchanged.

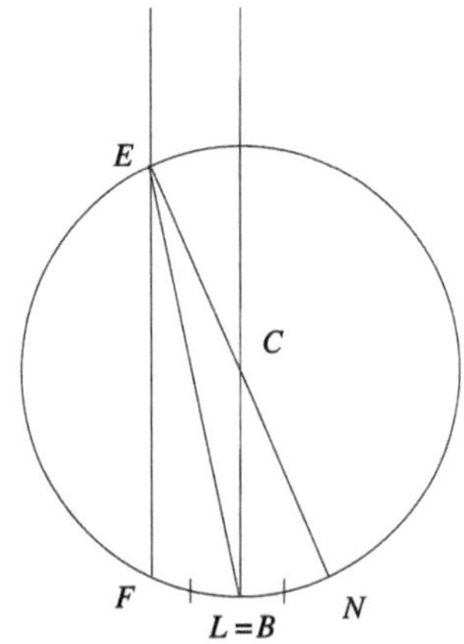

Fig. 8.3 Theorem XII. Demonstration, first part
∢*FEN* is the inclination angle and ∢*FEL* is the fraction angle. If *EL* arrives at *B*, then ∢*FEN*:∢*FEL*=2 (or <2 if *EL* intersects *AB*) which is impossible because ∢*FEN*:∢*FEL* = i:f = 8:3

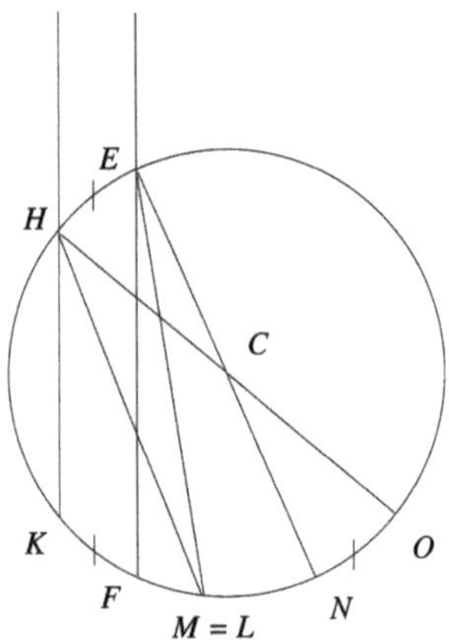

Fig. 8.4 Theorem XII. Demonstration, second part
If *HM* arrives at *L* then ∢*MHO*−∢*LEN* = arc *NO* and ∢*KHM*−∢*FEM*= arc *KF*. But arc *NO* is the difference between inclination angles, arc *KF* the difference between fraction angles and *KF*=*HE*=*NO*. So, the difference between inclination angles is equal to the difference between fraction angles which is impossibile, because $i : i' = f : f'$ and $i > f$, $i' > f'$

two simulations (Fig. 8.5) in order to plot the paths of light rays inside the glass sphere, according to Maurolico's law and according to Snell's law.[29]

Another application of the refraction law is found in theorem XIX: "If a ray passing through a transparent sphere parallel to the diameter intersects the diameter outside the sphere, then the ratio between the radius of the sphere and the distance of the point of intersection is always greater than the ratio between the fraction

[29] I used GeoGebra's tool "trace" to obtain the simulations. For this software, released under Non-commercial License Agreement, see https://www.geogebra.org/.

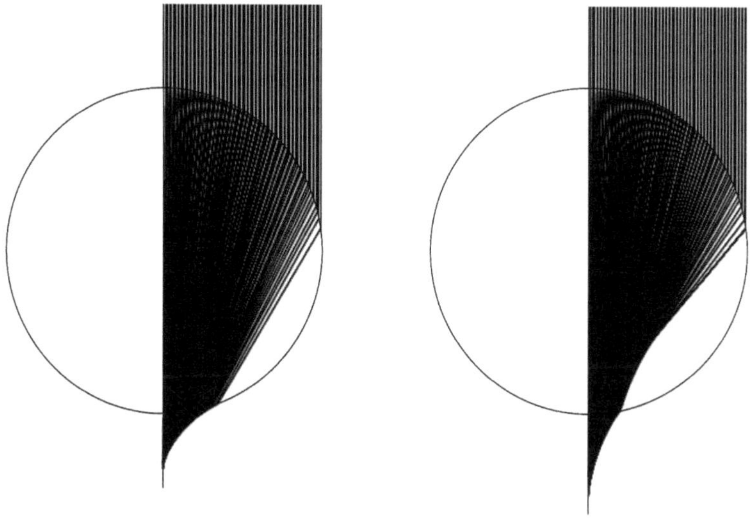

Fig. 8.5 Simulations with GeoGebra. Maurolico's law, $i:f = 8:3$ (*left*). Sine law, index = 1.5 (*right*)

angle and the angle between the refracted ray and the line perpendicular (to the surface of the sphere)".[30] In the corollary to this theorem, the corollary of theorem X is applied finding that the numerical value of the ratio between the distance of the point of intersection *DE* and the radius *FD* is always less than 5:3. This gives a correct value for the focal point, assuming a refractive index of 1.6. In fact, from theorem XIX we have (see Fig. 8.6):

$$FD : DE > \sphericalangle BEF : \sphericalangle BFE.$$

Corollary of theorem X implies $i{:}f = 8{:}3$, or (after some manipulation) $f{:}i{-}f = 3{:}5$. Since $\sphericalangle BEF$ and $\sphericalangle BFE$ are respectively the fraction angle and the difference between inclination angle and fraction angle, we obtain:

$$FD : DE > 3 : 5 \Rightarrow DE : FD < 5 : 3 \Rightarrow DE < \frac{5}{3} FD$$

This value corresponds exactly to the focal point for a refractive index of 1.6.[31]

[30] "Radio intra sphaeram pellucidam diametro sibi parallelo extra congrediente, sphaerae semidiameter ad distantiam congressus maiorem semper sortietur proportionem, quam angulus fractionis ad angulum, qui sub perpendiculari et fracto comprehenditur" (Diaph. § 74, p. 43). See Fig. 8.6: $FD{:}DE > \sphericalangle BEF{:}\sphericalangle BFE$.

[31] Maurolico, although drawing a sphere, is dealing with a hemisphere. The ray *AB* is parallel to the diameter *CD* and is not refracted at the surface *GAC*, as entering an hemisphere. From lens-maker's formula we obtain, for an hemisphere of radius *R* in air:

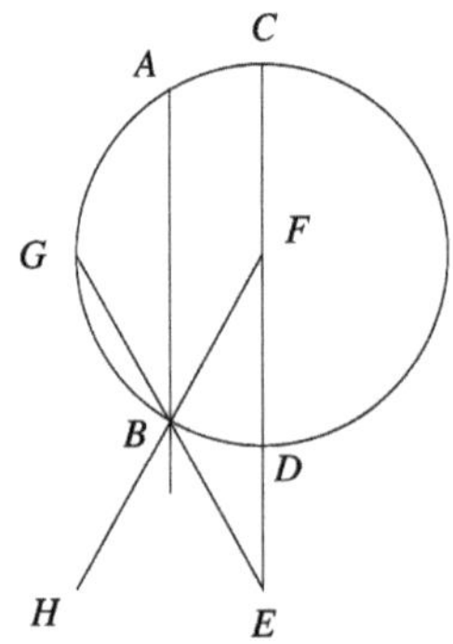

Fig. 8.6 Theorem XIX and its corollary
Based on the principle of reversibility (explicitly stated in assumption 4 of *Photismi*), Maurolico refers to the ray inside the sphere as refracted ray (*BA*) and the ray outside as the incident ray (*EB*). With this convention:
$\sphericalangle EBH = \sphericalangle FBG$ is the incidence angle (i);
$\sphericalangle ABG = \sphericalangle BEF$ is the deviated angle (f, contained by *BA* and *BG*);
$\sphericalangle FBG - \sphericalangle ABG = \sphericalangle FBA = \sphericalangle BFE$ is the angle included by the refracted ray (*BA*) and the perpendicular (*FB*) (which equals $i-f$)

8.6 The Maurolician Approach: A Success or a Failure?

Maurolico was completely confident that the theory of proportion could explain nature. In the case of refraction, he probably did not choose the correct magnitude to geometrise among those at his disposal: angle, arc, or segment etc. However the relationship between his law and its consequences is consistent. He demonstrated, *more geometrico*, theorems on light behaviour in glass spheres. And finally, in the third part of *Diaphana* (not examined here, but which deserves examination), he applied these consequences to the study of vision problems, giving a good account of them.

Maurolician approach represented also a failure: the law is experimentally incorrect, although it can be plausible for small angles. Moreover, Maurolico did not consider other media and his experimental apparatus is not clear.

To sum up, Maurolico gained some successes, because he gave attention to the need for a general law dealing with refraction and he showed how to derive a physical model from the Euclidean theory of proportion. His most important contribution

---

$$\frac{n}{p}+\frac{1}{q}+\frac{1-n}{R}$$

with $p = +\infty$ for parallel incident rays. So, in theorem XIX, substituting $R = FD = -3$ and $q = ED = 5$ (according to the Cartesian sign convention) n=8:5 is found, the refractive index implied by the corollary to Maurolico's law. The lens-maker's formula is obtained in paraxial approximation, in which case a law of proportionality between incidence and refraction angles holds, since $\theta \approx \sin\theta$ for θ sufficiently small.

is the use of a geometrical model for describing the behaviour of light with the deduction of direct consequences from the model itself. The model, for example, enabled him to find the distance of the focal point.

8.6.1 *The Circulation of Maurolico's Optical Studies*

Optical works were written in 1521–1523 but printed only in 1611. Consequently, there arises the question of a circulation of Maurolico's optical studies before their printing. In the preface of 1611 edition (Maurolico 1611, p. 4), there is a passage that opens, indeed, a possibility of an "underground" circulation:

> For since manuscript copies of his book were in hands of various people and although they were filled with errors and lacked the notes of Clavius, there were men who were striving to write upon optics; and it cannot be doubted that these men would have transferred everything from the flourishing gardens of Maurolycus to the barren fields of their own books, without any mention of his name.[32]

Is this just a rhetoric expedient of the printer to sell many copies of the book, or is there something that could strengthen the assertion of plagiarism? In fact, there is something more: *Photismi* and *Diaphana* were delivered by Maurolico to Christopher Clavius, the Jesuit mathematician of the Collegio Romano, on the occasion of Clavius's stay at Messina in 1574, as evidenced by Maurolico's nephew, Francesco Jr. Barone della Foresta, who writes in the biography (Maurolico Jr. 1613, p. 50):

> to whom [Clavius] he [Maurolico] gave the original of *Photismi* and *Diaphana*, divided into three books, in order to print them in Rome.[33]

Thus Maurolico intended to print the works on optics, with the intervention and the help of Clavius. However, the publishing project soon ran aground and no more news about the texts are found for several years. The reasons for Clavius's lack of interest are also unclear; it was probably due more to incidental reasons (pressing educational and scientific commitments) than to his actual lack of interest. Maurolico's works were certainly not forgotten; on the contrary, it is very likely that they continued to circulate in the Jesuit environment and arose interest, as witnessed by the passage quoted above, from the 1611 edition.[34]

---

[32] "Nam cum horum librorum manuscripta exemplaria in multorum manibus essent, quamquam mendis referta et sine Clavii notis, nec deessent qui Perspectivam suis scriptis illustrare molirentur, dubitandum non erat, quin hi ad suorum librorum campos exornandos e florentibus Maurolyci hortis, suppresso auctoris nomina, omnia transtulissent". English translation is from (Crew 1940, p. 3).

[33] "Vennevi etiando il Clavio giesuita celeberrimo nella professione e strinsero insieme ...molta famigliarità; a cui egli assegnò nel prender congedo l'originale de *Fotismi* e dell'opera delli *Diafani*, distinto in tre libri, affine di farglili stampar in Roma".

[34] There are more elements on the circulation of Maurolico's optical works, see (Bellé 2006b) for an analysis of the publication of Maurolico's optical works by C. Clavius and G. G. Staserius (S.J.

8.7 G. B. Della Porta and Optics: An Overview

The importance of Della Porta's *Magia naturalis* for the European culture of the first half of the XVII century is acknowledged and discussed in many articles.[35] The focus here is on the optical contributions, from the first edition of *Magia naturalis* (1558) until *De refractione* (1593).

Between 1593 and 1605, Della Porta published four scientific works that aimed to establish himself as a devotee to restore ancient science, departing from the themes and spirit of *Magia naturalis*: *De Refractione* (1593); *Pneumaticorum libri tres* (1601); *Elementa curvilineorum* (1601); *Magnae Constructionis liber primus* (1605).

The publication of these texts marked a new period in Della Porta's career. It is possible to identify some similarities in their development and approach. They make explicit reference to classical works written by renowned Greek mathematicians: Hero for *Pneumatica*, Ptolemy for *Almagest* and Euclid for *Elementa*, *Optics* and *Catoptrics*. Della Porta is now a translator, one of those who recover ancient mathematics, referring to past authorities, and importantly no more a weird scholar showing miracles of nature. But he is more than a translator: he adds his own researches with his own demonstrations. He discusses optics, astronomy, pneumatics and pure mathematics. *De refractione* and *Pneumatica* share also their origin: they both derive from extended sections in the second edition of *Magia naturalis*. These topics are covered in books XVII, XVIII and XIX of 1589 edition.[36]

So in sum, Della Porta enlarged a nucleus of speculations from *Magia* by reshaping his reasoning in a more "scientific" manner. He gave explanations with demonstrations, inserted citations and passages taken from the texts of Greek science, but he also proposed new interpretations based on his own observations.

8.7.1 Evolution in Approach to Optics

For a reassessment of Della Porta's optics it is necessary to make an analysis of the evolution of his interest in this field. It is important to establish what Della Porta wrote and when, in order to put in a larger context the ideas and results contained in *De refractione*.

Della Porta's interest in optics dates at the very beginning of his career. In the first edition of *Magia naturalis* (1558), book IV is already devoted to optics. Optics

1565–1635), or the Introduction to our edition in volume 10 of *Edizione Nazionale*, www.maurolico.it/Maurolico/sezione.html?path=10.

35 For example (Gardair 1990; Beaulieu 1990; Battafarano 1990; Eamon,1994; Balbiani 1999,2001).

36 For optics see next section. For *Pneumatica* refer to the Introduction to the critical edition of the text (Trabucco 2008), published in *Edizione nazionale delle Opere di Giovan Battista della Porta*, where also the critical edition of *De refractione* will be published. For further information, http://www.lincei.it/files/convegni/710_invito.pdf.

will be for Della Porta a lifelong interest, until the very end of his life. Just a few days before his death he said: "the challenge of the telescope will kill me, being the most difficult challenge I have ever faced".[37]

For what concerns optics, his studies could be divided into three periods:

1. Youth period, corresponding to the composition of the first edition of *Magia naturalis*. Della Porta's interest is in *demiranda experimenta*, the exceptional and wonderful effects that optics can produce. In this period a systematic and consistent approach is lacking.
2. Central period, corresponding to the second edition of *Magia naturalis* and *De refractione*. The analysis and description of experiments became more accurate and organised, in view of the renewed edition of *Magia*. The coherent study of other authors produced a more systematic approach. The general structure of *De refractione* will resemble classical Greek geometrical works or their Renaissance reconstructions.
3. Last period, characterised by the relationship with Cesi and the Accademia dei Lincei. Della Porta tried, unsuccessfully, to complete the numerous challenges of his life. The composition of *De telescopio* dates back to this period. It is a work composed to explain how telescope works by reshaping materials taken from *De refractione*.[38] To this period belongs also the (in)famous letter in which Della Porta ascribes to himself the invention of the telescope (Galilei 1934, vol. X, p. 252, letter n. 230).

Let's concentrate on the second period, starting with some references to the first, leaving aside the third period of his life.[39]

8.7.2 *Optics in the Two Editions of* Magia naturalis

The four books of *Magia naturalis, sive De miraculis rerum naturalium* are printed in 1558 by a young Della Porta. The fourth book, devoted mainly to optics, opens with a preface where the content and the approach are described: a sequence of *demiranda experimenta*, like mirrors reversing images, showing foot upwards and so on. However, optics pertains to the domain of geometry.[40]

---

[37] Stigliola to Cesi (10 April 1615): "che visitandolo [Della Porta] due giorni inanzi che si mettesse a letto in quest'ultima sua infermità, mi disse che l'impresa del Telescopio l'haveva ammazzato, essendo, come egli diceva, la piú difficile impresa et la piú ardua di quante mai havesse pigliato". Letter n. 393 in (Gabrieli 1996).

[38] A transcription of the autograph manuscript, Roma, Biblioteca dell'Accademia nazionale dei Lincei e Corsiniana, *Archivio Linceo* XIV, is found in (Naldoni and Ronchi 1962).

[39] Something on this last period, with an analysis of *De telescopio* can be found in Borrelli (2017).

[40] "Geometriae pars ...idest perspectiva appellatur, ad oculum pertinens, multa enim huiusmodi demiranda facit experimenta, ...rectius speculum inspectando pedes sursum, caput deorsum videntur, et caetera, uti latius in sequentibus perspiciuntur" (Della Porta 1558, p. 141).

Della Porta is not interested in discussing the explanation of these phenomena, referring the reader to a long tradition before him: Archimedes (maybe in connection with the burning mirror), *Optics* and *Catoptrics* by Euclid, Ptolemy and Witelo. Della Porta's aim is not to study the laws of optics in order to explain the effects, but to show how it is possible to increase these effects.[41] The most influential topics are probably the *camera obscura* and the *specillum*.[42] The description of *camera obscura* starts with the most simple case, but shortly after Della Porta adds a more complex structure: a concave spherical mirror,[43] whose role is to set upright the images in the *camera obscura*, otherwise reversed. Already in the 1558, moreover, Della Porta proposed an analogy between the eye and the *camera obscura*, giving a clear function to the "parva magnae spherae portio, ultimo locata oculi", that is the retina, although its function is restricted to the reflection of images and not to the perception itself.[44]

The case of *specillum* is rather different. In the first edition of *Magia*, an analysis of lens properties and a description of their effects are lacking: lenses are used only for multiplying images. By contrast, in 1589 edition (and even more in *De refractione*) the chapters devoted to the lenses are the most influential and innovative. In 1558, in fact, Della Porta is not yet aware of the true capabilities of lenses as vision instrument: for him lenses are just toys.[45] There follows a quite wide and consistent discussion about mirrors and their effects. This part represents the kind of speculation Della Porta is interested in: there is a mirror in which images are seen flying and some mirrors which show images hanging down.[46]

An analysis of the concave spherical mirror is found in chapter XIV, where a rule for determining the position and the type of images is also given. Della Porta thinks

[41] "Reddere tamen eorum causas, alienum puto, cum disciplina ea affatim satisfaciat, multique eas reddiderint; quas qui praeoptarit, consulat Archimedem Syracusanum, Euclidis Opticam, et Catoptricam, Ptolemaeum, Vitellionem et reliquos, quorum multa desumemus, multaque denuo excogitata adiiciemus, ut hinc quisque possit in infinitum ea propagare; ut in omnibus evenit experientiis" (Della Porta 1558, p. 141).

[42] "Quomodo in tenebris ea conspicias, quae foris a sole illustrantur et cum suis coloribus", in chap. II and "Quomodo res multiplicata videri possit", in chap. IV. Della Porta, in fact, will deal again with these topics in the 1589 edition, expanding his analysis. Moreover they represent the most important contribution of Della Porta to the history of optics, as testified, for example, by the references found in Kepler's *Paralipomena* or by the claimed precedence in the invention of the telescope.

[43] "Non quod disgregando dissipet, sed colligendo uniat", according to Della Porta's description, (Della Porta 1558, p. 143).

[44] "Intromittendi dirimitur quaestio sic agitata … intromittitur enim idolum per pupillam fenestrae instar, vicemque obtinet speculi parva magnae spherae portio, ultimo locata oculi; quod si quis distantiam mensuraverit, centri loco fiet visus" (Della Porta 1558, p. 143).

[45] "Inter ludos qui circunferuntur, non parum iucunditatis est specillum; instrumentum vitreum illud, quod oculis apponimus, ut commodius aliquid intueamur" (Della Porta 1558, p. 145).

[46] Many descriptions are taken directly from Ptolemy's *De speculis*. This work was ascribed to Ptolemy in that period (Ptolemy 1518). W. Schmidt and L. Nix considered it as written by Hero (Hero 1900, pp. 303–315). More recently (Jones 2001) questioned both of these attributions and published it as Pseudo-Ptolemy.

that the point which determines the variation in dimension and position of images is the center of the circumference.[47] The part devoted to optics ends with the burning mirror (chapter XV) and with other optical devices for producing fire.[48]

Between the two editions of *Magia*, Della Porta's interests in optics continued to grow, as testified by a group of his letters written around 1580 in Venice, during the period under Cardinal Luigi d'Este.[49] Della Porta was involved in building a parabolic mirror in Venice in collaboration with Giacomo Contarini, one of the Commissioners of the Arsenal.[50] Della Porta worked also on an "occhiale" and searched for some "fugacciette di vetro" (lenses) in Murano.[51]

So, it does not happen accidentally that, for what concerns optics, the additions and variations between the two editions of *Magia* are mostly found in the part on burning mirror and lenses. In the 1589 edition, however, another important improvement concerns the concave spherical mirror. In the chapter IV (corresponding to chapter XIV of 1558 edition) Della Porta introduces the *point of inversion* and the *cathetus rule* to localise the image. At the very beginning of the chapter, the part devoted to the inversion point is found (Della Porta 1589, p. 264): "Punctum inversionis imaginum in concavo speculo cognoscere". Della Porta describes a procedure very similiar to that found in 1558 edition: place the mirror facing the sun and look for the point where the rays intersect.[52] After the determination of the inversion point a series of effects related to it are described.[53] Concerning the camera obscura,

[47] "It will be easy to understand all the variations [of the concave mirror], if you find the center of the concave semicircle, being that all the things are regulated and known from it. If you want to see the head upside-down, you have to put your head outside the center of the mirror and you will see the head upside-down and the foot up", chap. XIV (Della Porta 1558, pp. 150–151): "Centro concavi hemicycli iam invento, diversitates omnes cognoscere facile erit cum ex eo omnia regulentur et noscantur. Si vis ergo inversum spectare caput, extra speculi centrum caput habeto, et inversum statim inspicies, et pedes sursum". In the 1589 edition, instead, he will define rightly this point as *punctum inversionis*, which corresponds to our focal point. However Della Porta did not give any quantitative information on its position, but only suggested that it could be found by observing the figure made by solar rays, reflected in the mirror. A mention to this method is already found in the 1558 at the end of chap. XIV: "Splendentis Solis radis speculum opponat, et fomitem vel accendibilem rem centri loco collocet, quod accedendo et removendo invenies lucisque conus ostendit et inducit illico flammam".

[48] For example, a "phiala" filled with water or a round crystal. (Della Porta 1558, pp. 151–152).

[49] On this letters in connection with optics see (Bellé 2005, p. 9). All the letters to Cardinal d'Este are found in (Campori 1872).

[50] Giacomo Contarini is a well known figure of the period, also in connection with Galilei. See, for example, the letters between Contarini and Galileo in (Galilei 1934, vol. X, pp. 52, 55, 57).

[51] "Incontrai il S. Leonardo col quale di nuovo siamo stati a Murano al Ballarino, se havesse fatto le fugacciette di vetro per gli occhiali e la fontana", 10th December 1580, (Campori 1872, p. 22).

[52] "Oppone speculum soli et ubi radios coire videbis, scito ibi esse punctum inversionis".

[53] "Si extra punctum inversionis fueris, inversum carnes caput". On the discovery of *punctum inversionis* by Ausonio see (Dupré 2005). Later on in the same chapter IV Della Porta wrongly considered the center of the circumference like the inversion point (as he had done in 1558): "The reason is that rays that go from the center to the circumference after reflection become parallel, since parallel rays, after reflection on the circumference, intersect themselves in the *center* [our italics]". "Ratio est, quod radii a centro ad circumferentiam reflectuntur parelleli [*sic*] quum paralleli ad

the most important innovation is the idea to put a lens in its hole.[54] This has some influence also on the analogy between camera obscura and eye, because in the eye a lens is present: the crystalline. The eye description is accordingly revised.[55]

The text is still a *Magia*, so there are many examples of strange and astounding effects even in optics. However a more refined theoretical structure is found. Della Porta has become a well-versed scholar: in 1558 he considered "alienum" to explain the causes and reasons of the strange and wonderful phenomena. Even more, the astonishment itself is derived from the lack of explanation. In the 1589 edition, however, his attitude is changed a bit so as to insert some additional results, like the aforesaid inversion point or the cathetus rule. This is testified also by the preface to book XVII, where Della Porta assigns to optics the role of confirming the correctness of mathematical demonstration.[56]

In 1589 *Magia naturalis* Della Porta refers to another work on optics written by himself, with the phrase "ut in nostris opticis".[57] The reference is made to *De refractione*, published only four years later. So let us turn to this work.

8.7.3 De Refractione

De refractione optices parte libri IX was published in Naples in 1593. It is a text different from the two *Magiae* in content and style. First of all, it is completely devoted to a specific phenomenon: refraction. *De refractione* moreover is presented as a text of classical geometry: it opens with definitions and postulates and then proceeds with theorems proved from the suppositions.[58] Also the relation between mathematics and optics and the drive for composing the work are quite different. Optics is born from a mixture of mathematics and natural philosophy and makes use

centrum concurrant" (Della Porta 1589, p. 264). This mistake bears witness to a textual stratification in the evolution of the second edition and to its strong connection with the first one.

[54] "Si crystallinam lentem foramini apponens, iam iam omnia clariora cernes", chap. VI (Della Porta 1589, p. 266).

[55] "Intromittitur idolum per pupillam, fenestrae *foraminis* instar, vicemque obtinet *tabulae crystallinae* spherae portio, *in medio* oculi locata", in italics differences with the 1558 version.

[56] "Quid enim ingeniosus excogitari potuit, ut imaginariis animi conceptionibus certissima experimenta subsequerentur? Et mathematicarum demonstrationum vera fides etiam ocularibus experimentis comprobaretur?" (Della Porta 1589, p. 259).

[57] The statement "in nostris opticis" is found six times in the whole book XVII of the 1589 *Magia*: chap. VI, on the relation between the eye and the camera obscura: "in nostris opticis fusius declaratum est", p. 267; chap. X, *de crystallinae lentis effectibus*, two instances: "sed de his plenius in nostris opticis" and "rationes in opticis reddidimus", p. 269; chap. XIV, on spherical burning mirror: "latius de ea in opticis", p. 271; chap. XIX, on burning by refraction: "rationem reddemus in opticis", p. 277; chap. XX, on a specific kind of mirror: "in opticis diximus", p. 278.

[58] Book III, devoted to the description of the eye and book IV, on the theory of vision, make exception to this "mathematical structure".

of mathematical demonstrations. Della Porta was compelled to write anew the work by the inadequate treatments of previous work on refraction.[59]

De refractione is a rather long text, divided into nine books and dealing with many aspects of refraction: from rainbow to binocular vision, from eye structure to lenses and optical illusions. In this article the focus will be on the relation between book VIII, on lenses, and book II devoted to establish how light passes through glass spheres. Della Porta tried to justify lenses' properties on the basis of geometrical constructions and mathematical demonstrations, although the deductive structure is not so convincing.[60]

8.8 Della Porta on Refraction

The first book starts with seven definitions, rather standard in the period,[61] devoted to the decription of the technical terms adopted in the text: *linea incidens*, *linea refracta*, *angulus incidentiae* and *angulus refractionis*.[62]

The first propositions explain the refraction using a sort of experimental instrument, made up by the *mater* and the alidade of an astrolabe (Fig. 8.7). This instrument should be partly immersed in water and the observer uses the alidade for checking the path of solar ray in the passage between air and water. It should be linked with the device described by Ptolemy, Alhacen and Witelo.

In the third proposition Della Porta establishes the law of reversibility of optical path.[63] In this proposition, moreover, one of the very few "quantitative statement" is found. As an application of the instrument, Della Porta writes (1593, p. 12) that the solar ray, falling with an angle of 45 degrees, will get closer to the perpendicular, obtaining 35 degrees.[64] This values is not found in Ptolemy's (and consequently

[59] "[Optica] ex mathematices et naturalis philosophiae mixtura orta est. ...Hinc geometricarum speculationum veritas innotescit, nam quae geometria fingit et speculatur, ipsa explicat et in lucem revocat". "Causa, quae nos de refractione ad scribendum impulit, fuit, quod videremus ea de re maiores nostros breviter, ac satis oscitanter perscripsisse, ut vestigia potius quaedam, quam artis fundamenta iecisse viderentur" (Della Porta 1593, pp. 5–6).

[60] The ultimate goal is to apply what is deduced for the glass sphere to the eye: "Haec enim pila oculi specimen refert, nec vera a nobis loca visionum in oculis decerni possunt", bk. II, preface, (Della Porta 1593, p. 35). I shall not deal in this paper with this rather problematic issue.

[61] Compare for example, the definitions found in the book X of Witelo's *Perspectiva*, (Witelo 1572, p. 404); Della Porta's seven definitions correspond to definition 1, 4, 3, 5, 6, 8, 10 in Witelo's text, with some differences.

[62] As noted before (p. 7, n. 14), also Della Porta calls *angulus refractionis* the angle of deviation, not "our" refraction angle.

[63] "Ingredientes egredientesque semper aequalis in refractione reperiuntur" (Della Porta 1593, p. 10).

[64] See Fig. 8.7. "Sit Sol in astrolabio *D*, in medio quartae, ad gradus 45, non veniet ad *E*, ubi est etiam 45 gradus, sed accedendo ad *HBI* in *F*, erit 35 gradus. At existens Sol in *F*, [ubi] 35, non veniet ad *G*, ubi etiam 35 sed ad *D*, ubi gradus 45. Remanent ergo anguli in medio aequales *GBD* et *EBF* ex lineis se decussantibus orti, ad invicem aequales".

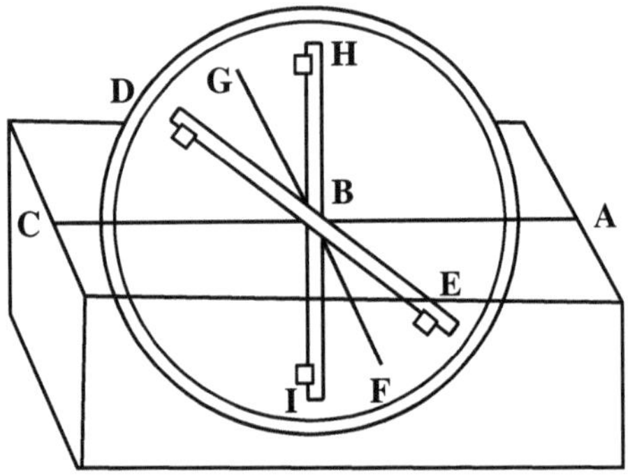

Fig. 8.7 Apparatus described by Della Porta

Witelo's) tables.[65] Della Porta probably obtained the value from some experiments with the apparatus described.[66]

The following propositions, quite traditional in optics, are devoted to establish and apply the rules for the localisation of images.[67] In proposition 8, for example, Della Porta (1593, pp. 16–17) tries to correct the proposition 14 of book X of Witelo (1572, p. 415).[68]

8.8.1 *Della Porta on Glass Spheres*

The second book deals with the study of light rays passing through glass spheres. The key idea is to exploit the analogy between reflection in spherical mirror and refraction in glass sphere. This part of Della Porta's optics arose interest in the last years and some recent papers have analysed it in details.[69]

The laws of refraction proposed in the second book are quite problematic and have deserved negative and positive treatments in literature. But, despite the

[65] Ptolemy and Witelo, in fact, give measures of angles multiples of ten, from ten degrees up to eighty degrees, see (Smith 1996, p. 233) and (Grant 1974, p. 425).

[66] There is an error of about three degrees (in water, refractive index 1.33, the angles of refraction is 32 degrees). Three degrees is a quite reasonable value, taking into account the structure used by Della Porta. An error of three degrees, if the alidade was 50 cm long, corresponds to a displacement of about 2.5 cm of the ray under water, compensated by the alidade's thickness. Della Porta, in this case, seems to consider the angle between refracted ray and perpendicular line, that is the refraction angle.

[67] Location of images in Della Porta is analysed in details in (Borelli 2014, pp. 53–57). As said before, cathetus rule is already present in the second edition of *Magia* (see p. 15).

[68] Here is Della Porta enunciation: "Something under water, seen by refraction, will be seen more elevated if it is more distant from the eye". For a deep examination of this proposition, see (Borelli 2014, p. 50).

[69] I shall rely on and follow, in particular, (Dupré 2006, pp. 33–36), (Smith, 2010, xciii-xcv, figures pp. 436–439), (Zik and Hon 2012, pp. 445–456) and (Goulding 2014, pp. 145–147 and 171–174).

judgment on the laws itself, I would like to focus on its use by Della Porta to prove the subsequent propositions. For our convenience, let us explain with an example. Maurolico's law is wrong, since angles of incidence and refraction are not proportional one to another, given that their sines are proportional. But, despite this error, Maurolico demonstrated in a perfect mathematical style some consequences of his law. Clearly, also the consequences do not fit exactly with the real behaviour of light. According to Maurolico's law, as seen before, parallel rays entering a glass sphere would not intersect each other inside the sphere, but actually they do. However, Maurolico's demonstration is mathematically correct. In another different world, where refraction were proportional to the angles (and not to their sines), refracted rays would not intersect each other inside a glass sphere, exactly as demonstrated by him.

Now the question is: did Della Porta make correct application of his own laws of refraction in order to find (or to justify) the behaviour of lenses? The function of mathematics can be twofold in regard to natural sciences: it can be used as a research tool in order to discover new relations in nature, as probably done by Maurolico; or it can simply justify—or even better, demonstrate—some properties discovered by experiences or trial and error method.

In proposition 1,II (see Fig. 8.8) Della Porta considers an incident ray *AB* cutting an arc *BC*, on the circumference *DBC* centered in *E*, equal to one sixth of the circumference.[70] In this case the angle is 60° and so the ray is reflected in *C*, end-point of the diameter parallel to incident ray. Della Porta goes on with other rays more or less distant from the vertical diameter. He considers only rays which cut arcs equal to the sides of the octagon and dodecagon but he gives a general rule at the end of 1,II (Della Porta 1593, pp. 40–41). Moreover, he proves that the reflected ray cannot

Fig. 8.8 Theorem 1, bk. II

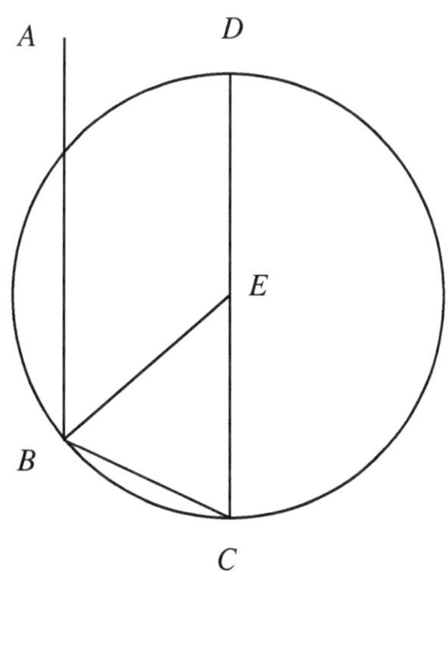

[70] Hereinafter propositions from *De refractione* are cited with their number followed by a Roman numeral referring to the book. I call this ray hexagonal, being that *BC* is equal to the side of inscribed hexagon.

ascend "ultra quartam diametri". This is an important result, because it can give a correct estimate for the focal point of a spherical mirror.

In proposition 2,II he applies what he found on reflection to refraction points. As D. C. Lindberg (1984, p. 143) wrote: "the transference of these same principles to refraction is, to say the least, problematic". In fact, Della Porta's reasoning is based on this analogy (Della Porta 1593, p. 43): "ubi in concavo ascendebant, hic descendunt, et ubi ibi descendebant, hic ascendunt". This means that, if the point of intersection between the *reflected* ray and the diameter parallel to incident ray goes up, nearer to the center of the circumference, then the corresponding *refracted* ray, inside a glass sphere, moves in the opposite direction: it goes down. According to him, since the reflected ray, as it goes nearer the diameter, ascends towards the center of E, the refracted ray will descends of the same amount, *outside* the sphere.[71] The hexagonal ray, which cuts the end of the diameter in reflection, intersects the diameter in the same point also in refraction. Moreover, from the condition on the reflected ray, which cannot ascend more than a quarter of the diameter, one can deduce (but Della Porta did not) that the refracted ray cannot intersect the diameter more distant than a quarter of the diameter. The focal point of a sphere would be determined as $R/2$, where R is the radius of the sphere.[72]

The first problem in applying this rule is that the ray should be refracted again coming out from the spherical surface. Della Porta makes no mention of this question. Probably this rule should be considered a "holistic" law, which does not follow all the passages but only gives the comprehensive result. To be clearer: we should apply refraction law twice, that is, firstly when the ray enters the sphere and secondly when the ray exits.[73] Della Porta looks at the entering ray, then he applies the law and finds the point where the ray, *outside* the sphere, cuts the diameter parallel to the entering ray. What happens inside the sphere is not specified. In fact, the figure illustrating theorem 2,II shows no bending in exiting rays. A quite strange, but however tenable, principle. Robert Goulding (2014, p. 147) had defined Della Porta rule a "black box":

> Della Porta, on the other hand, treated his instruments as a kind of "black box": light entered one side, something happened to it in the middle, and it exited the other to reach the point predicted by his model.

Probably this rule derives from the fact that it is easier to observe what happens outside a sphere. Everything inside a sphere is blurred and confused. Whereas Della Porta was able to see parallel solar rays outside the sphere and could find where these rays intersect each other on the other side, what happens inside was

---

[71] Also Kepler in his treatment of law of refraction made a large use of analogy between refraction and reflection. See (Malet 1990, pp. 281–282).

[72] I emphasise that *we* can deduce, but Della Porta did not. A similiar situation is found in (Zik and Hon 2012, p. 455) where, applying proposition 5,II, the authors (not Della Porta) found an estimate of 2.4 times the radius for the focal point of an hemisphere.

[73] I add that some authors give a different interpretation of the theorem: in their view there is no need to assume that II,2 deals with double refraction. See for example (Smith 2015, pp. 345–349), (Zik and Hon 2012, pp. 449–450) and Borrelli (2014, p. 52).

regrettably unclear to him. After this observation Della Porta found an empirical rule and tried to find a geometrical justification, on the basis of the analogy with spherical mirrors.

The result obtained for the sphere is then applied with some difficulties to the case of plane-concave and plane-convex lenses. For concave lenses Della Porta (1593, p. 43) made use of the analogy between concave and convex: "concavum esse convexum contarium" and so, rays in the concave hemisphere will be refracted in the opposite direction compared to the convex one. The final section of the book deals with location of images, by the use of cathetus rule applied to spheres. This topic will be resumed in book VIII, where Della Porta treats lenses, and, in particular, the position and the dimension of the images of an object seen through the lens.

8.8.2 Della Porta on Lenses: Book VIII

In book VIII, Della Porta tries to apply the results of book II on glass spheres to lenses. He takes under scrutiny biconvex and biconcave lenses with such extension and analysis of many cases that no one made before. He concludes with an explanation of their use for correction of far- and short- sightedness. Correct and complete results in lens theory are found, but their demonstrations are, to say the least, problematic. In proposition 2 there is a rule to find inversion point, while propositions 3–5 analyse what happens when the object or the eye are in the inversion point.[74] The following group (propp. 6–9) are devoted to convex lenses and finally, propp. 15–16 to concave lenses. All the theorems share the same structure and are divided into two parts: tracing of the path of the rays, by the laws of refraction of book II; finding position of image by cathetus rule. I focus here on ray tracing, trying to find a connection with book II.

Let us examine what is found in book II regarding this topic. The enunciation of proposition 16,II is very clear: "If the object is beyond the point of inversion it will be seen reversed".[75] As said before, Della Porta introduces inversion point also for lenses and glass spheres, a good idea.[76] Unfortunately, in book II there is no definition of point of inversion (as required in a mathematical treatise) but neither a practical procedure to find this point is proposed to the reader. Inversion point is central in his theory: "every kind of variation of images depends on it".[77] In the demonstration of the theorem (see Fig. 8.9) the inversion point is *L*, the object seen *GF* and the eye is placed in *A*.

There are two issues in this demonstration that deserve our close inspection: the application of refraction laws and the inversion point. The path of refracted rays

[74] As will be seen, in *De refractione* Della Porta introduces inversion point also for the lenses.

[75] "Magnitudo extra punctum inversionis posita conversa videbitur" (Della Porta 1593, p. 56).

[76] For another analysis of inversion point, with a different point of view, see Zik and Hon (2017).

[77] "Punctus inversionis maxime in eiusmodi apparentiis vestigandus est, nam ex eo omnes varietates nascuntur" (Della Porta 1593, p. 56).

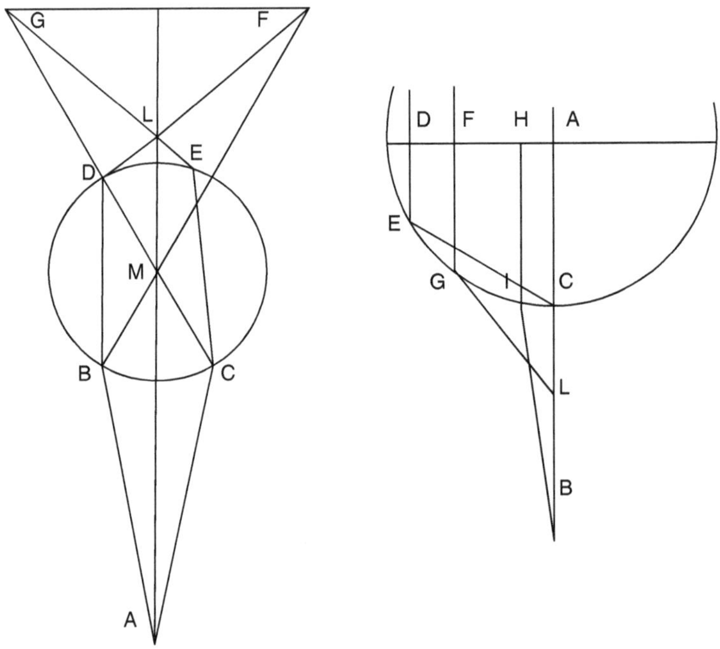

Fig. 8.9 Inversion point in theorem 16, bk. II (*left*) and theorem 2, bk. VIII (*right*)

through a glass sphere is determined in proposition 2 of the same book (it does not matter at this moment whether this determination is right or wrong). In our interpretation (see p. 22) this proposition deals with rays passing through a complete sphere, so it should be enough to determine the path also in this case. On the contrary, two references are made: the first one for the entering ray (proposition 2,II) and the second for the exiting ray (proposition 5,II).[78] In all the propositions of book II the same happens: proposition 2 applies only to the first half surface encountered by light, while proposition 5 applies to the other half surface of the sphere. Moreover, propositions 2 and 5 deal with the same case: the intersection of parallel rays with the parallel diameter. Although this aspect is lacking in proposition 16, it is possible to determine, by applying proposition 2 in this case, only where the ray *GE* will intersect a parallel diameter on the other side of the sphere.[79] But this is not how the propositions are applied in 16,II. They are used only in a qualitative manner: the rays will be bent towards the perpendicular line when they enter on one side and come out on the other from the sphere. The quantitative aspect of preceding propositions is completely ignored here. Della Porta, for example, did not specify if the

[78] "Ex *D* refrangetur ad *B* per secundam huius et egrediendo veniet ad oculum *A* per quintam" (Della Porta 1593, p. 57). Proposition 5,II deals with parallel rays exiting from a convex hemisphere (Della Porta 1593, p. 47).

[79] The same holds for proposition 5, from which, it is possible to find only where the ray *EC* will intersect a parallel diameter.

ray *GE* (or the ray *EC*, inside the sphere) is a hexagonal ray or the point where the ray cuts the circumference, information needed to apply correctly theorem 2,II.

As for the inversion point, after proposition 16,II there are other cases where the object and the eye are located near or far from the sphere, and the outcome is correctly described according to the inversion point. But the point of inversion is never determined numerically, though possible with just a bit of (more or less explicit) calculation. To our surprise, the point is known only *a posteriori*, that is by moving the eye. For example, summing up the content of propositions 17 and 18 of book II (Della Porta 1593, pp. 58–60) one can conclude that the object lies beyond inversion point if, moving the eye far away, the object could be seen inverted; otherwise, if the object is always seen erected—no matter how far the eye is moved—then the object lies between the point of inversion and the sphere.[80] These propositions do not give a rule for finding the inversion point but instead give a rule to determine whether an object lies beyond or before inversion point. Where the point of inversion is located can only be determined by trials and errors, placing the object nearer to the sphere, or farther from the sphere, and moving the eye away to find if the object is seen reversed or not.

Proposition 2,VIII: "To find point of inversion in convex lenses" should be connected with 16,II.[81] In this case, Della Porta gives a rule to determine inversion point: the point of intersection of the refracted rays is the inversion point. He refers to proposition 3 of book II.[82] The reference to prop. 3,II should be instead to prop. 5,II. In the Fig. 8.9, rays *DE*, *FG* and *HI* are parallel to the radius *AC* inside the hemisphere *CIGE*, so it should be applied proposition 5,II. But looking at the demonstration a description similar to that of 2,II is found. In fact the hexagonal ray is refracted at the end of the diameter (point *C*) and the rays "quae sunt supra *E*" are not relevant since they are refracted inside the sphere.[83] This phenomenon is only described in 2,II, when the pentagonal ray cuts the diameter "supra finem diametri", inside the sphere. It is not clear what Della Porta refers to, there is no coherence between book II and book VIII for what concerns inversion point and both descriptions are defective. In book II a quantitative application of preceding propositions is lacking. In book VIII an attempt is found but it is not clear how to apply Della Porta's "laws of refraction".

Browsing book VIII another similar situation is found. Proposition 15,VIII (Della Porta 1593, p. 185) deals with bi-concave lens. As usual, the proposition expresses rightly the phenomenon and the enunciation is clear and satisfactory: "In

[80] Prop. 17: "If the object and the eye are near the sphere, the object will be seen upright, but if the eye goes back, the object is seen inverted". Prop. 18: "If the object is near the sphere, within inversion point, it is seen upright, whether the eye is far or near to the sphere".

[81] "In specillis convexis inversionis punctum invenire" (Della Porta 1593, p. 175). Proposition 16,II deals with spheres but Della Porta considered lenses as composed by two spherical surfaces.

[82] "Puncta concursus refractarum linearum, tam convexi quam concavi corporis, sunt puncta inversionum" (Della Porta 1593, p. 176). There is no attempt to find an estimate of the position of inversion point, although using 5,II it would be possible, see (Zik and Hon 2012) and n. 71 on p. 25.

[83] "Nam intra crystalli corpus refranguntur" (Della Porta 1593, p. 176).

concave lenses the object is always seen smaller". Della Porta (Fig. 8.10) traces ray *AI*, refracted in *M* for *tertium quinti* [correct: 3,II] and the refracted in *G* for *quartum supradicti nostri* [4,II]. After this ray tracing, the cathetus rule is applied.[84]

The problem arises because this demonstration does not represent a correct and coherent application of the laws of refraction, as exposed in book II. The correct references are made to the relevant propositions[85]: 3,II deals with ray entering a concave surface (in this case *EHID*) and 4,II with a ray exiting from a concave surface (*ELMD*).[86] But their application fails. In the drawing, the rays inside the lens are bent on the wrong side, but in the text the description is correct according to (Borelli 2014, p. 57). But I think that the real problem is another, more general and more damaging for Della Porta's mathematical structure. In the application of the quantitative results of propositions of book II, it is essential to specify if rays *AI*, *BH* and *IM*, *HL* cut the surface of the sphere at a distance equal to the side of the hexagon, or of the octagon or somewhere else. The position of the point of intersection with the surface is necessary to find the path of refracted ray. Otherwise it is impossible to establish if the ray follows *that* path or a slightly different one. Della Porta, instead, in every instance, uses the propositions of book II only in a qualitative way, discarding completely the quantitative description of *his own* laws of refraction. To demonstrate, as he did, the propositions of the last part of book II and all the theorems of book VIII, it was sufficient a qualitative law like the following one: "light

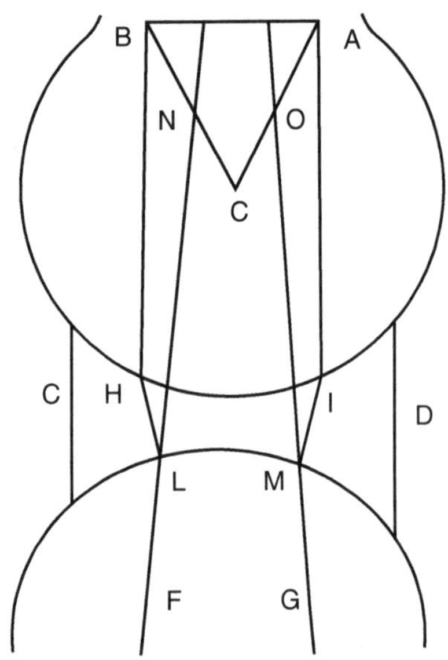

Fig. 8.10 Theorem 15, bk. VIII

[84] For the problems in applying cathetus rule in this situation see (Borelli 2014, p. 59).

[85] In the text, reference is made to "tertium quinti", with a misprint in book number. Proposition 3,V deals with parallel lines which appear intersecting (Della Porta 1593, p. 112). The correct book reference is to book II.

[86] In the figure point *E* is wrongly indicated as *C*.

passing from a less dense medium to a denser one is bent towards the perpendicular line. When light follows the inverse path, the opposite occurs". But how much are rays bent? Propositions 2–5 of book II are quantitative laws and they give an estimate of the bending of the rays. The treatment of lenses in book VIII, instead, is only a qualitative description of the behaviour of rays which is presented in "pretended" geometrical manner. Book VIII shows no application of the quantitative results found in book II. He proposed in book VIII a thorough and correct analysis of lens properties, useful for the later researchers. The demonstrations of these properties, from a geometrical point of view, are not satisfactory. Probably he arrived at those results by trials and experiments, definitely not by applying a mathematical theory to natural phenomena.

8.9 Della Porta: Success or Failure?

As in the case of Maurolico, also in Della Porta the dark and bright sides are found. His optical theory is an advancement for what concerns at least three aspects:

1. He gave a good and systematic account of convex and concave lenses, analysed as composed by two spherical surfaces, each one submitted to its own refraction.
2. He proposed the study of lenses according to the inversion point, already introduced for reflection in spherical mirror. He discussed also a way to find it, although this issue is more problematic.
3. Della Porta tried to distinguish between what an optical element does and what the eye sees; a very important and fruitful approach to optical problems, in general, not only in lens theory.
4. Finally he demonstrated that analogy could be a research tool as effective as mathematics to investigate nature.[87]

For materials analysed above, there are also some controversial points. The most important is probably the lack of mathematical rigour and the complete absence of a quantitative description for lenses. In dealing with lenses, he proposed only a practical law, although presented in a pretended geometrical way.

From our study of *De refractione*, it is evident that Della Porta's approach is not a mathematical one, in terms of a quantitative interpretation of physical phenomena. He tried however to give to his work a shape similar to the classical works, but his approach is still lacking the necessary internal coherence.[88] This is also testified by the huge number of misprints (and in some cases, even more serious errors) in the

[87] After all, nowadays the study of geometrical optics starts with spherical mirrors and then proceeds to lenses, exploiting the analogy between these two optical artifacts.

[88] This trait is peculiar of Della Porta "mathematician", as happens in *Elementa curvilineorum libri III* where, according to (Napolitani 1990, p. 156) Della Porta is a researcher fond not of the truth of his results but of the shock they can produce. A more detailed analysis of this aspect can be

text and in the figures illustrating it, which makes some passages unintelligible. Despite this, Della Porta gave to *De refractione* a structural clearness and he succeeded in presenting his results in a geometrical manner, not rigorous though.

8.10 Conclusions

Two tendencies can be identified in the study of optical phenomena during the sixteenth century: one is for the usage of experiments and observations in nature; the other is for much interest in constructing a geometrical model of natural phenomena. Whatever path is followed, something will be excluded from the theoretical model. If the geometrical way is chosen, it will be impossible to describe all the empirical facts, because of the interpretation of the model. On the other side, when a scholar wants to describe every effect, or at least a great part of them, it will probably be impossible (or very hard) to demonstrate everything from a mathematical point of view in a persuasive manner. Science makes progress, on the one hand, by taking under scrutiny the facts that are outside the mathematical model, and, on the other, by trying to mathematise models which describe most of the facts. In other words, science progresses by attaining a proper balance between theory and experience.

Maurolico's and Della Porta's optical studies show that these two approaches were not totally separated: they are both present in each author. Nevertheless, Maurolico was more inclined towards a mathematical approach. He produced a law of refraction perfectly deduced from his hypothesis, but without enough experiments and observation (only one case is recorded). On the other hand, Della Porta reported a lot of observations. He was unable to put them in a coherent and strong geometrical frame. Maurolico and Della Porta obtained some interesting results, but they lacked the complete vision of the problems concerned with refraction. The optical "revolution" of the first half of the seventeenth century by scholars like Kepler and Descartes was built on the achievements of authors like Della Porta and Maurolico. Their approaches are both needed to find (and demonstrate) a tenable law of refraction. Scholars following Maurolico and Della Porta will make the most of the results and the methods developed in the sixteenth century.

References

Balbiani, Laura. 1999. La ricezione della Magia Naturalis di Giovan Battista Della Porta. Cultura e scienza dall'Italia all'Europa. *Bruniana & Campanelliana* 5: 277–303.

found in (Gavagna and Leone 2005), the critical edition of the text, published in *Edizione nazionale delle Opere di Giovan Battista della Porta*.

———. 2001. *La Magia Naturalis di Giovan Battista Della Porta. Lingua, cultura e scienza in Europa all'inizio dell'età moderna*. Bern/Frankfurt/M.: Peter Lang.

Battafarano, Italo Michele. 1990. Spee – Harsdörffer – Knorr von Rosenroth. La ricezione di Giovan Battista Della Porta in Germania dalle polemiche demonologiche alla genesi del linguaggio scientifico tedesco. In *Giovan Battista Della Porta nell'Europa del suo tempo*, ed. Maurizio Torrini, 311–336. Napoli: Guida editori.

Beaulieu, Armand. 1990. L'influence de Della Porta sur la physique en France au XVIIe siècle. In *Giovan Battista Della Porta nell'Europa del suo tempo*, ed. Maurizio Torrini, 291–310. Napoli: Guida editori.

Bellé, Riccardo. 2005. *Il De refractione di G.B. Della Porta. Edizione critica*. PhD Thesis (unpublished).

———. 2006a. Il corpus ottico mauroliciano: origini e sviluppo. *Nuncius* 21: 7–29.

———. 2006b. I gesuiti e la pubblicazione dell'ottica di Francesco Maurolico. *Bollettino di storia delle scienze matematiche* 26: 211–243.

Borrelli, Arianna. 2014. Thinking with optical objects: Glass spheres, lenses and refraction in Giovan Battista Della Porta's optical writing. *Journal of Early Modern Studies* 3: 39–61.

———. 2017. Optical diagrams as "paper tools": Della Porta's analysis of biconvex lenses from *De refractione* to *De telescopio*. In *The Optics of Giambattista Della Porta (ca. 1535–1615): A reassessment*, ed. Arianna Borrelli, Giora Hon, and Yaakov Zik. Cham. Springer.

Campanus, Johannes. 1546. *Euclidis ...Elementorum geometricorum libri V cum expositione ... Campani*. Basel: per Johannem Hervagium.

Campori, Giuseppe. 1872. Giovanni Battista Della Porta e il cardinale Luigi d'Este. *Atti e memorie delle Reali Deputazioni di Storia Patria delle provincie modenesi* 6: 1–27.

Cardanus, Hieronymus. 1570. *Opus nouvum de proportionibus numerorum, motuum, ponderum, sonorum, aliarumque rerum mensurandarum*. Basel: Ex officina Henricpetrina.

Clavius, Christopher. 1574. *Euclidis Elementorum libri XV ...Auctore Christophoro Clavio*. Roma: apud Vincentium Accoltum.

Clubb, Louise George. 1965. *Giambattista Della Porta. Dramatist*. Princeton: Princeton University Press.

Crew, Henry. 1940. *The Photismi de lumine of Maurolycus. A Chapter in Late Medieval Optics*. Macmillan: New York.

Della Porta, Giambattista. 1558. *Magiae naturalis, sive De miraculis rerum naturalium libri IIII*. Napoli: apud Matthiam Cancer.

———. 1589. *Magiae naturalis libri XX*. Napoli: apud Horatium Salvianum.

———. 1593. *De refractione optices parte libri novem*. Napoli: apud Io. Iacobum Carlinum & Antonium Pacem.

Dupré, Sven. 2005. Ausonio's Mirrors and Galileo's Lenses: The Telescope and Sixteenth-Century Practical Optical Knowledge. *Galilaeana* 2: 145–180.

———. 2006. Visualisation in Renaissance Optics: The Function of Geometrical Diagrams and Pictures in the Transmission of Practical Knowledge. In *Transmitting Knowledge. Words, Images, and Instruments in Early Modern Europe*, ed. Sachiko Kusukawa and Ian Maclean, 11–39. Oxford: Oxford University Press.

Eamon, William. 1994. *Science and the Secrets of Nature. Books of Secrets in Medieval and Early Modern Culture*. Princeton University Press: Princeton.

Gabrieli, Giuseppe. 1996. *Il carteggio linceo*. Roma: Accademia Nazionale dei Lincei.

Galilei, Galileo. 1934. *Edizione nazionale delle opere*. Vol. X. Firenze: G. Barbera editore.

Gardair, Jean Michel. 1990. L'immagine di Della Porta in Francia. In *Giovan Battista Della Porta nell'Europa del suo tempo*, ed. Maurizio Torrini, 273–290. Napoli: Guida editori.

Gavagna, Veronica, and Carlotta Leone. 2005. *Elementorum curvilineorum libri tres*. Napoli: Edizioni scientifiche italiane.

Giusti, Enrico. 1986. Ricerche Galileiane: il *De motu aequabili* come modello della teoria delle proporzioni. *Bollettino di storia delle scienze matematiche* 6: 89–108.

———. 1993. *Euclides Reformatus. La teoria delle proporzioni nella scuola galileiana*. Bollati Boringhieri: Torino.
———. 2001. Maurolico et Archimède: sources et datation du 1er livre du 'De momentibus aequalibus'. In *Medieval and Classical Traditions and the Renaissance of Physico-Mathematical Sciences in the 16th Century. Proceedings of the XXth International Congress of History of Science*, ed. Pier Daniele Napolitani and Pierre Souffrin, 33–40. Turnhout: Brepols.
Goulding, Robert. 2014. Thomas Harriot's Optics, Between Experiment and Imagination: The Case of Mr Bulkeley's Glass. *Archive for History of Exact Sciences* 68: 137–178.
Grant, Edward. 1974. *A Source Book in Medieval Science*. Cambridge: Harvard University Press.
Heath, Thomas. 1956. *The Thirteen Books of the Elements*. New York: Dover Publications.
Hero of Alexandria. 1900. *Opera quae supersunt Omnia. Mechanica et catoptrica*, ed. L. Nix et W. Schmidt. Leipzig: B. G. Teubner.
Jones, Alexander. 2001. Pseudo-Ptolemy De Speculis. *SCIAMVS* 2: 145–186.
Lindberg, David C. 1984. Optics in 16th-Century Italy. In *Novità celesti e crisi del sapere: Atti del Convegno Internazionale di Studi Galileiani*, ed. Paolo Galluzzi, 131–148. Firenze: Giunti Barbera.
Malet, Antoni. 1990. Gregoire, Descartes, Kepler and the Law of Refraction. *Archives Internationales d'Histoire des Sciences* 40: 278–304.
Maurolico, Francesco. 1528. *Grammaticorum rudimentorum libelli sex*. Messina: Petrutius Spira.
———. 1968. *Prologi sive sermones quidam de divisione artium, de quantitate, de proportione*, ed. Graziano Bellifemine. Molfetta.
———. 1611. *Photismi de lumine & umbra ad perspectivam & radiorum incidentiam facientes. Diaphanorum partes ...tres ...Problemata ad perspectivam & iridem pertinentia*. Napoli: ex typographia T. Longi.
Maurolico, Francesco Jr. 1613. *Vita dell'Abbate del Parto D. Francesco Maurolico*. Messina: per Pietro Brea.
Naldoni, Maria Amalia, and Vasco Ronchi. 1962. *De telescopio*. Firenze: Leo S. Olschki.
Napolitani, Pier Daniele. 1988. La geometrizzazione della realtà fisica: il peso specifico in Ghetaldi e Galileo. *Bollettino di storia delle scienze matematiche* 8: 139–237.
———. 1990. La matematica nell'opera di Giovan Battista Della Porta. In *Giovan Battista Della Porta nell'Europa del suo tempo*, ed. Maurizio Torrini, 113–166. Napoli: Guida editori.
Ptolemy, Claudius. 1518. *Liber de speculis*. Venezia: Luca Antonio Giunta.
Ronchi, Vasco. 1968. *Scritti di ottica. Tito Lucrezio Caro, Leonardo Da Vinci, G. Rucellai, G. Fracastoro, G. Cardano, D. Barbaro, F. Maurolico, G. B. Della Porta, G. Galilei, F. Sizi, E. Torricelli, F. M. Grimaldi, G. B. Amici*. Milano: Il Polifilo.
Rose, Paul Lawrence. 1975. *The Italian Renaissance of Mathematics. Studies on humanists and mathematicians from Petrarch to Galileo*. Librairie Droz: Genève.
Smith, Mark A. 1996. *Ptolemy's Theory of Visual Perception*. Philadelphia: The American Philosophical Society.
———. 2010. *Alhacen on Refraction*. Vol. 1. Philadelphia: The American Philosophical Society.
———. 2015. *From Sight to Light. The Passage from Ancient to Modern Optics*. Chicago: The University of Chicago Press.
Trabucco, Oreste. 2008. *Pneumaticorum libri tres*. Napoli: Edizioni scientifiche italiane.
Witelo. 1572. *Opticae Thesaurus*. Basel: per Episcopios.
Zamberti, Bartholomeus. 1505. *Euclidis ...Elementorum geometricorum libri XV*. Venezia: In aedibus Ioannis Tacuini.
Zik, Yaakov, and Giora Hon. 2012. Magnification: How to Turn a Spyglass into an Astronomical Telescope. *Archive for History of the Exact Sciences* 66: 439–446.
———. 2017. Giambattista Della Porta: A magician or an optician?. In *The Optics of Giambattista Della Porta (ca. 1535–1615): A reassessment*, ed. Arianna Borrelli, Giora Hon, and Yaakov Zik. Cham. Springer.

Chapter 9
Conclusion: A Reassessment

Yaakov Zik and Giora Hon

Della Porta's optical works have mostly received mixed if not negative appraisals in the historiography of early modern optics. Given the puzzling lapses and logical inconsistencies in his optical works, one cannot be surprised that students of the history of optics have repeatedly depicted him as either ignorant of known optical principles or unable to see the flaws of his own reasoning. The articles comprising this volume redress the epistemological background, merits, and failures of Della Porta's scholarship in a critical yet balanced approach. We commence the concluding chapter with a survey of the context in which Della Porta worked so that his genuine contribution to the practice of optical science (*optica scientia*) will be at hand.

Della Porta's life reflects the scientific culture in Renaissance Naples. Humanism was a distinctive characteristic of Della Porta's natural philosophy. However, it did not mean here a revival of classical literature and philosophy; rather, it is the belief that natural phenomena were primarily defined by how they were perceived, felt and acted upon by those who experienced them (Borrelli; Eamon). The way Della Porta experienced natural phenomena coheres with the metaphor of theater of experiments that characterized Renaissance Neapolitan scientific culture. Della Porta's experiments rehearsed the principles of natural magic which taught that nature is alive and sentient, filled with powerful occult forces to which only the learned magus had access. Natural magic made perfect sense to Neapolitans and Della Porta offered them tools for mining its secrets (Eamon; Dijksterhuis). In fact, these principles were not restricted to Renaissance Naples; they were also dominant in the tradition of natural magic in the Low Countries. The many editions of *Magia Naturalis*, published all over Europe, attest to the influence Della Porta had well beyond the Neapolitan scientific culture (Dijksterhuis; Bellé).

From the second half of the thirteenth century onwards, the term *perspectiva* referred explicitly to the study of geometrical optics, with the goal of explaining

Y. Zik (✉) • G. Hon
Department of Philosophy, University of Haifa, 31905 Haifa, Israel
e-mail: zikya@013.net; hon@research.haifa.ac.il

A. Borrelli et al. (eds.), *The Optics of Giambattista Della Porta (ca. 1535–1615): A Reassessment*, Archimedes 44, DOI 10.1007/978-3-319-50215-1_9

visual perception. By Della Porta's time the term took on the current meaning, with its reference to artistic perspective as a means of depicting scenes in three-dimensional space on a plane surface. Della Porta, however, had in mind the optical and not the artistic meaning. He had a critical attitude towards past optical authorities in general and Alhacen and his three Perspectivist followers, Witelo, Roger Bacon, and John Pecham, in particular. The Perspectivist tradition had manifested several issues which Della Porta addressed: (1) quantitative relationship between the angles of incidence and refraction; (2) optical magnification; (3) visual perception beyond the boundaries of the pyramid of vision (i.e., peripheral vision), and (4) image formation. The Perspectivists studied primarily visual perception, so that optical images were always conceived as constructs seen by an observer. In this context, no distinction was possible between "virtual", "subjective" images and "real", "objective" ones, which can be physically projected on a screen (Smith). The emergence of this distinction can be seen as marking a critical turning point in the history of optics, and Della Porta's work was seminal in this respect.

Towards the end of the sixteenth century the idea of different kinds of images was extraordinary, if not preposterous. The Perspectivists offered a qualitative account of the path of a visual ray refracting through a single interface; in their view the study of refraction through a second interface would contribute nothing new to optical knowledge. Optical elements and lenses in particular cause objects to appear displaced or distorted due to magnification or diminution; indeed, objects sometimes appear inverted. In other words, optical elements caused modifications if not distortion of natural vision. For these reasons the study of optical elements was considered to be an exercise in practical applications rather than a study of theoretical implications. The Perspectivists concentrated on explanatory implications, and had little or no interest at all in practical applications (Smith; Borrelli; Bellé).

Della Porta's optical study marked a fundamental change in the understanding of optics in the second half of the sixteenth century. His early writings provide evidence of systematically empirical studies of reflective/refractive properties of mirrors, glass spheres, and lenses. These studies had merits as well as problems (Borrelli; Zik & Hon). Della Porta's account of sight in *De Refractione* followed the classical and medieval tradition, particularly that of the Perspectivists. But the goal he set himself was operative, that is, to reproduce the effects in an attempt to understand them and ultimately reconceptualize their principles. This was a different epistemic goal than the formulation of general geometric laws of light and vision. It could not be attained by complying with considerations related to subjective vision. As far as visual perception was at stake, the point of inversion——where the image, produced by concave mirrors and convex lenses, is seen magnified, turned upside down and blurred——was a useful, yet an unreliable reference point. The cathetus line, dropped orthogonally from a point on an object and passing through the center of curvature to the place where the eye is located, was another reference, in this instance, a line. Its primary function was to help locate in three-dimensional space the apparent position of an image seen by the eye through a mirror or a refracting element. Della Porta was willing to part ways with his classical and medieval antecedents at several junctures and in the process sought new analytic directions that were however problematic. Just how problematic the results could be is manifest in his accounts of peripheral vision, diplopia, and image location (Smith; Borrelli; Bellé).

Della Porta's programmatic approach to optics and his studies of optical phenomena turned out to defy the Perspectivist tradition. The projection of a real image on a screen is one such example. The capacity of a small aperture to project inverted pictures into a darkened room (i.e., *camera obscura*) was well known. But Della Porta was the first to realize the function of the pupil, as an aperture, in the transmission of images into the eye. Like Girolamo Cardano, Della Porta suggested applying a convex lens at the aperture, but he improved the optical setup by placing a concave spherical mirror beyond the aperture. Adjusting and directing the optical elements towards a screen, he successfully projected an upright, enhanced and vivid picture. Another example was the correction of visual disorders by convex and concave lenses, and the adjustment of pair of lenses as an optical device for improving vision by showing far and near objects, more distinct at the eye. Della Porta acknowledged the following problematic optical issues: (1) Why do light rays reflecting/refracting by/through a spherical interface end up at different points along the optical axis (i.e., spherical aberration)? (2) Does sight depend on the intromission or extramission of rays? (3) What are the causes of size variations of the pupil in accordance with light intensity? And (4) What are the causes of "ghost images" (i.e., disturbing appearances) which hinder the use of plane and spherical thick back-coated mirrors? Della Porta developed novel explanations and genuine ways to resolve these issues (Bellé; Borrelli; Heeffer; Smith).

What were then Della Porta's contributions to the practice of optical science:

1. In *Magia Naturalis* and *De Refractione*, Della Porta investigated optical phenomena as a kind of magic stripped however of its occult and demonic aspects in a procedure we have come to call experimental science. His goal was to obtain a rigorous geometrical procedure for image construction and, in the case of refraction in glass sphere, he actually delivered it (Borrelli; Bellé; Smith; Zik & Hon). In the case of spherical lenses, he provided a strict rule for constructing the refracted ray. In his later, unpublished writing, he sought to formulate a fitting version of the cathetus rule (Borrelli). However, he was apparently not able to bring that work to completion.
2. The point of inversion and the point of burning (i.e., where the optical element ignites fire) were considered to occupy the same location at the center of the optical element. Della Porta discovered that the position of the point of inversion and that of the point of burning occupy different spatial locations. He associated the point of inversion with a locus where the image is seen magnified, inverted, and blurred——a matter of visual perception. He defined the point of burning as a physical, optical position associated with a geometrical point (i.e., focal point) in which the converging rays ignite fire. This is an important conceptual move. In his demonstrations of optical phenomena, Della Porta distinguished perceptual considerations from purely geometrical-physical aspects (Borrelli; Smith; Zik & Hon).
3. Della Porta's demonstrations were dependent on knowledge of the exact proportions of the radius curvature of the mirrors he used. This requirement motivated him to investigate the working of reflecting elements. His theory of reflection

delineates a physical law governing the path of rays reflected by concave spherical mirrors. Excluding considerations of image formation and visual perception, Della Porta developed a mathematical formulation which defines a point, that is, the point of reflection (*punctus reflexionis*), where the reflected rays meet the optical axis of any given concave spherical mirror at a distance equals to half of its radius curvature (Zik & Hon).

4. Della Porta's study of reflection in spherical mirrors and refraction in glass spheres was based on a systematic tracing technique for illustrating the path of the rays in terms of their directions and the angles by which they converge or diverge. This was a rigorous geometrical procedure of image construction which, in the case of glass spheres, yielded quantitative predictions. The importance of seeking quantitative results cannot be exaggerated——it is a characteristic of Della Porta's innovative approach to optical science (Bellé; Heeffer; Zik & Hon).[1]

In conclusion Della Porta's optical studies present two principal avenues of research: (1) an account of phenomena associated with sight and how does vision occur (Smith); and, (2) practical rules of manipulating optical elements for reproducing specific visual effects (Borrelli; Zik & Hon). In the former Della Porta complied with the Perspectivist notions of lenticular mode of vision; however, in the latter he broke away from this tradition. His distinction between the visual point of inversion and the point of burning marked a fundamental change. While the point of inversion is not a measurable point as it shifts continuously and could be determined at several different locations, the point of burning is a geometrical point determined by the physical properties of the optical element (Bellé; Zik & Hon). Della Porta realized that, without exact knowledge of the physical properties of the optical elements he used, he could not provide rules on how to obtain the required visual effects. To generate this knowledge Della Porta linked experimental practice with geometrical analysis, and this combination in some cases yielded a qualitatively, and eventually also quantitatively, successful procedure for matching observed phenomena.

Della Porta formulated a mathematical scheme based on inference from experiments which provided cogent approximation for tracing reflected and refracted rays. This approximation was the only available quantitative theory of refraction at the time. It was a theory which skilled opticians, using a straightedge, compass, and table of chords, could easily apply in their practice. The fact that Della Porta made a mathematical scheme to be an approximation of a physical law does not mean that the scheme cannot be modified through subsequent scientific inquiries. Physical law is not absolute; sometimes it can have exceptions, be proven wrong, modified, and evolved over time (Bellé; Heeffer). Given the context in which Della Porta was

[1] For rays incident below the chord of 16 sided polygon the quantitative values appear in good agreement with experiment. For further discussion of Della Porta's search for quantitative results, see Y. Zik and G. Hon. 2012. Magnification: How to turn a spyglass into an astronomical telescope. *Archive for History of Exact Sciences* 66: 439–464.

working, it is not productive to consider his theory of refraction a failure. Instead, like Kepler, Harriot, and Descartes, one may regard Della Porta's approximation as an intermediate phase——the best he could achieve——in the pursuit for better understanding of the working of refractive elements (Heeffer). Della Porta contributed fundamentally towards the technological management of sets of optical elements. This approach was most useful in the development of the telescope where the critical aspect is not perception but rather ratio of spatial angles.

The essays comprising this volume show how much work remains to be done. Della Porta's optical studies contained significant contribution to the understanding of the functioning of optical elements. This understanding may facilitate further studies aimed at redressing Della Porta's optics and the impact he had on contemporaries and successors. One such example presented in this volume is the study of optical diagrams as "paper tools" (Borrelli; Bellé; Heeffer); it could serve different epistemic functions, thus suggesting a powerful heuristics for further investigations.

FSC
www.fsc.org
MIX
Papier aus verantwortungsvollen Quellen
Paper from responsible sources
FSC® C105338

If you have any concerns about our products,
you can contact us on
ProductSafety@springernature.com

In case Publisher is established outside the EU,
the EU authorized representative is:
Springer Nature Customer Service Center GmbH
Europaplatz 3, 69115 Heidelberg, Germany

Printed by Libri Plureos GmbH
in Hamburg, Germany